A Geography of Islands

Islands have always fascinated people. They often seem remote and mysterious, set between the continents on which most people live. Indeed, many people choose islands for their perfect holiday idyll. In practice, however, the everyday social and economic reality is often very different.

A Geography of Islands firstly examines the differing ways islands are formed. Despite the uniqueness of such islands in terms of shape, size, flora and fauna, and also their economic and developmental profiles, they all share certain characteristics and constraints imposed by their insularity. These present islands everywhere with a range of common problems. *A Geography of Islands* considers how their small scale, isolation, peripherality and often a lack of resources, has affected islands, in the present day and their past. It considers and discusses population issues, communications and services, island politics and new ways of making a living, especially tourism, found within contemporary island geography.

A Geography of Islands gives a comprehensive survey of 'islandness' and its defining features. Stephen A. Royle has visited and studied 320 islands in 50 countries in all the world's oceans. It is full of up-to-date global case studies, from Okinawa to Inishbofin, and Hawaii to Crete. In the final chapter, all the themes are brought together in a case study of the Atlantic island of St Helena. It is well illustrated with the author's own photographs and maps. This book will appeal to those studying islands as well as those with an interest in the topic, particularly those engaged in dealing with small island economies.

Stephen A. Royle is a Senior Lecturer in Geography at Queen's University, Belfast. He has researched on islands around the world, originally inspired by a holiday visit to Dursey Island, County Cork, Ireland.

Routledge Studies in Human Geography

This series provides a forum for innovative, vibrant, and critical debate within Human Geography. Titles will reflect the wealth of research which is taking place in this diverse and ever-expanding field.

Contributions will be drawn from the main sub-disciplines and from innovative areas of work which have no particular sub-disciplinary allegiances.

A Geography of Islands

Small island insularity

Stephen A. Royle

London and New York

To Katy, Alex and Ian,
for all those times I was away 'islanding'

First published 2001
by Routledge
11 New Fetter Lane, London EC4P 4EE

Simultaneously published in the USA and Canada
by Routledge
29 West 35th Street, New York, NY 10001

Routledge is an imprint of the Taylor & Francis Group

© 2001 Stephen A. Royle

Typeset in Galliard by Wearset, Boldon, Tyne and Wear
Printed and bound in Great Britain by MPG Books Ltd, Bodmin

British Library Cataloguing in Publication Data
A catalogue record for this book is available from the British Library

Library of Congress Cataloging in Publication Data
Royle, Stephen A.
    A geography of islands : small island insularity / Stephen A. Royle.
        p. cm.
    Simultaneously published in Canada.
    Includes bibliographical references and index.
    1. Islands – Maps. 2. Islands – Geography – Maps. 3. Islands –
Economic conditions – Maps. 4. Islands – Social aspects – Maps.
I. Title.

G1029 .R6 2001
912'.1942–dc21
                                                        00-045043

ISBN 1-857-28865-3 (hbk)

Contents

Figures

Tables

Acknowledgements

This book is the product of many years of visiting islands, initially those close at hand, more recently some far afield. I have had grants at various times from the Commonwealth Geographical Bureau, the Royal Society, the British Academy, the British Association for Canadian Studies, the Foundation for Canadian Studies and Queen's University Belfast. The Canadian government, through the High Commission in London, have made me grants for work in Canada (often with my colleagues Susan Hodgett and John Othick). The British Foreign and Commonwealth Office sent me to Ascension Island as part of a research team.

Some of the ideas here were developed in a summer school on small islands I gave for the Institute of Island Studies at the University of Prince Edward Island, Canada in 1991. I have been back often to the institute and value my contacts with Harry Baglole, the Director, also Ed MacDonald and Laurie Brinklow. Others I would like to thank include Basil George, former Chief Education Officer on St Helena and two South Atlantic stalwarts, Trevor Hearl and Alan Crawford. Russell King edited the first paper where I put some of the ideas into print and has remained in contact regarding island matters. Mark Hampton, secretary of the International Small Island Studies Association, was a fellow traveller to Ascension, as were John Christensen, Ben Catermoule and Nikola Farmer. John Connell of the University of Sydney is another islophile whose travels to islands, especially hot islands, inspired some of my own efforts. Klaus Dodds, Douglas Lockhart, Duncan MacGregor, David Phillips and Derek Scott are others who have helped me or worked with me on island matters. I was honoured to be asked to be the literary executor of the late Tony Cross whose researches on St Helena I doubt I have done justice to. Thanks to the generations of Queen's University Belfast geography staff and students who have put up with me on field trips to Cyprus, Mallorca and Malta. Finally, I would like to thank my family who have suffered my frequent absences on trips to islands or, perhaps worse, have been dragged along on insular family holidays, a.k.a. field trips. The most notable was that to the Outer Hebrides, not a cool place for Belfast schoolchildren to visit on holiday it seems, except in the climatological sense.

I would like to acknowledge the efforts of Andrew Mold and Ann Michael of Routledge and of two readers of the first draft of the book. The maps were drawn by Maura Pringle of the Queen's University School of Geography, to whom thanks. The photographs are all my own.

Gerald Freake of Newfoundland gave permission for the reproduction of Figure 6.2; Sandy Carruthers, Guardian News, for Figure 6.3; The *Gleaner* Company Ltd for Figure 7.6. Air Pacific gave permission for the reproduction of Figure 9.2.

1 Islands
Dreams and realities

Introduction

There are hundreds of thousands of islands. They are found in every sea and ocean and display considerable ranges of scale, resource availability, economic opportunities and levels of development (Figures 1.1–1.6). Yet it is this book's intention to show that, despite such wide distribution and variations, islands everywhere are subject to the impact of a common range of constraints imposed because of their very insularity. Such constraints – remoteness, smallness (absolute and/or relative), isolation, peripherality, etc. – can also affect, singly or together, certain mainland areas, but they are more notable in their effect on the bounded landmasses that are islands. The impact of insularity is, of course, usually more significant on small islands. Thus this book does not deal with islands of the scale of Great Britain, Honshu or Java. Instead, it focuses on smaller places, hence its subtitle: 'small island insularity'.

The effects of insularity are universal and, thus, this book has a global scope. However, it makes no attempt to be a catalogue of the world's islands. Instead it adopts a thematic approach. This opening chapter serves as a general intro-duction to the study of islands, dealing first with the pragmatic issues of where they may be found and how many there are, and what counts as an island anyway. It then considers the concept of islands and how this has entrained popular and artistic imagination, leading into a discussion of the 'island of dreams' conception. The chapter then returns to the real world by considering the use of islands and the concept of islandness in science, before concluding with a section on islands and reality – small island life is often at some remove from any expression of the island of dreams.

Most of the book is concerned with issues that fall within the scope of human geography, but it is vital that some consideration is given to the physical basis on which this human geography operates. Therefore Chapter 2 deals with the physical geography of islands. Their ranges of location and scale suggest that islands result from a great variety of processes of formation. Islands display a considerable range of habitats, too, although there are some commonalities brought about by insularity, such as a restricted variety of species. It is the author's contention that islands often share certain characteristics in social,

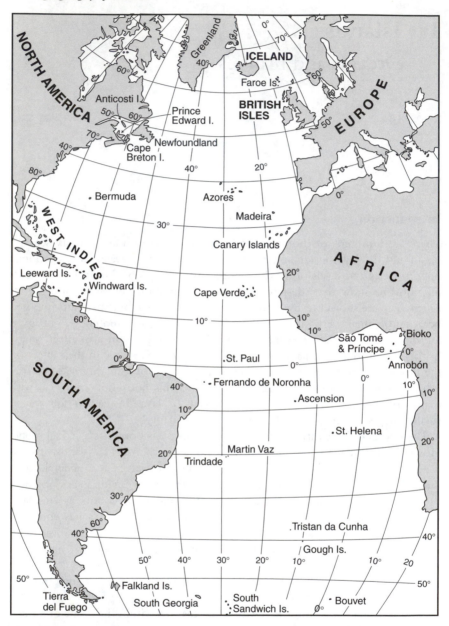

Figure 1.1 Islands of the Atlantic Ocean

economic and political spheres which are brought about simply by their being islands, and this conception is detailed in Chapter 3, which brings the introductory section of the book to an end.

Chapter 4 is a consideration of how islands were affected by their shared characteristics in historical periods when small islands were dependent on

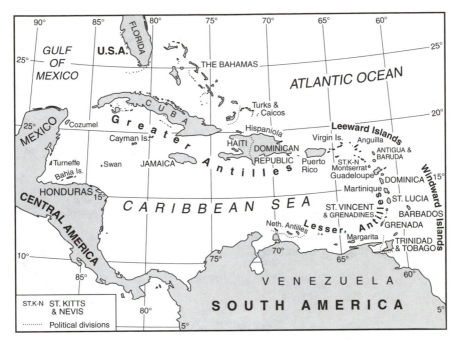

Figure 1.2 Islands of the Caribbean

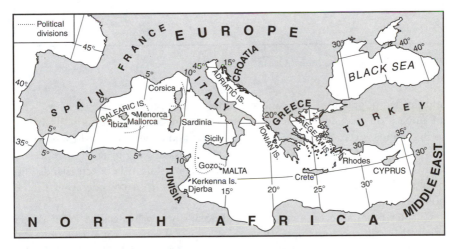

Figure 1.3 Islands of the Mediterranean

domestic resources to an extent not always seen today. Chapter 5 considers the way in which islands' peoples and population have been affected by their insularity. This may be in terms of out-migration, which has seen islands in the limiting case being completely abandoned. Elsewhere there is in-migration, which has bequeathed islands such as Mauritius, Trinidad and Fiji an ethnically diverse

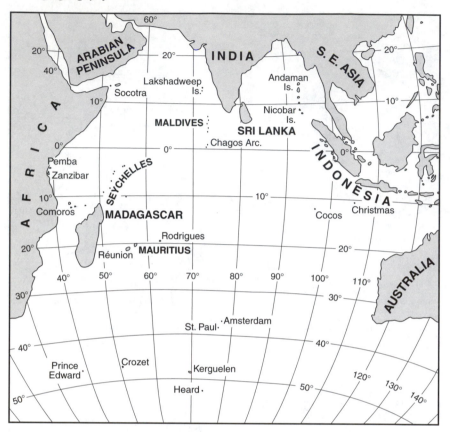

Figure 1.4 Islands of the Indian Ocean

and divided population. This can, sometimes, cause problems of otherness, with an impact on occasion upon social stability. One of the common problems facing islands is the small scale of their societies and economies, and these topics are explored in Chapter 6. Small places such as islands are usually powerless in political terms and the results of the unequal contestation between islands and outside forces are discussed in Chapter 7, along with a consideration of the politics of small islands.

Chapter 8 has an economic theme and details how a living is to be made on small islands in the contemporary world. Chapter 9 reports on a near universal development strategy, the adoption of tourism, part of which is predicated upon the innate romance and mystery of islands, the 'island of dreams' idea, reflecting back to this opening chapter.

Throughout the book, reference will be made to processes which will be shown to occur on islands throughout the world; the emphasis will be on the process and its application and not on the place. To conclude, it seemed logical to focus on one place instead, which exemplifies in depth all the processes

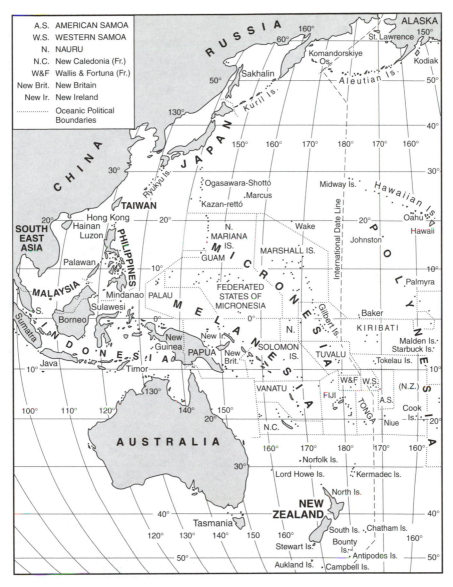

Figure 1.5 Islands of the western Pacific Ocean

operating. Therefore the final chapter, Chapter 10, is not simply a repetition in summary form of what has gone before. Instead it concludes this book's consideration of 'small island insularity' by detailing its impact upon one such small island, St Helena in the South Atlantic Ocean.

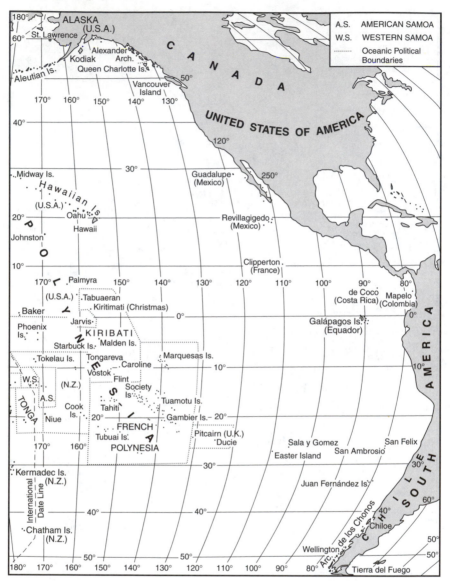

Figure 1.6 Islands of the eastern Pacific Ocean

Islands: where and how many?

Before the journey through the geography of the world's islands can begin, we should first consider the basic facts – where the islands are and how many there are of them. The book concentrates on offshore islands, but the issues of insularity discussed will also apply to continental islands situated in lakes and rivers (see, for example, Gutsche and Bisaillon, 1999).

It seems sure that the presence of all of the world's islands has now been noted. In earlier eras, voyages of discovery would sometimes end with intrepid mariners bringing back details of new islands. Many such 'discoveries' were that only in European terms, of course. This author attended the opening of a conference on islands held in Jamaica in 1992 when the (European) chair made a reference to it being 500 years since Columbus had 'discovered' the West Indies. The chair was promptly interrupted by a high-ranking official from the University of the West Indies who pointed out with vigour that there were many people fully aware of the existence of these islands long before Columbus. However, the European voyages of exploration included true discoveries of islands, which were uninhabited, some not known to humankind before. Some of these islands were of considerable significance, such as Mauritius and Réunion in the Indian Ocean.

Some discoveries were wrongly charted, some claims, through ignorance or mendacity, were not actually true and there were a number of completely apocryphal islands that appeared on maps and charts. These included the three Aurora Islands, supposedly between the Falkland Islands and South Georgia, first reported by the Spanish ship *Aurora* in 1762, found again by the same ship in 1774, by which time another Spanish captain, of the *San Miguel*, had charted their location as 52°37'S 47°49'W. In 1794 the Spanish corvette, *Atrevida*, even described the islands in detail, though the captain did not attempt a landing. Perhaps this was sensible, for it seems that trying to step ashore on the Auroras may have caused him to get wet, for they do not actually exist. James Weddell searched for them in 1820 as part of his Antarctic explorations. American sealer Benjamin Morrell also failed to find them in 1822, as indeed did everybody else who looked, including the armchair geographers, for these were the islands searched for by the eponymous character in Edgar Allan Poe's *The narrative of Arthur Gordon Pym of Nantucket* (1838). There was just one disbelieved sighting of five islands in the approximate area in 1856. The islands were no longer charted after about 1870. Ramsey postulates that they may have been icebergs improbably appearing as land in similar locations at a succession of sightings; that they may have been mismeasured sightings of the Shag Rocks, actually 53°S 43°W and much smaller, or perhaps they had existed but sank. He dismisses this last explanation as for the 'romantically minded' and has to conclude that the non-existence of the Aurora Islands is one of the 'unsolved mysteries of the sea' (1972, p. 80). However, volcanic islands have certainly disappeared within recent times, including Graham Island or Ferdinandea, south of Sicily, which having been seen in 10 BC, made a brief reappearance for a few months in 1831.

Ramsey goes on to detail the cases of a number of other either misplaced or entirely mythical islands, some given early cartographic recognition thanks to wishful thinking or superstition, many mythical islands being associated with enchantment or with the devil. Sometimes discoveries of islands were kept secret for strategic reasons, thus the Portuguese kept the existence of uninhabited St Helena in the Atlantic to themselves after finding it in 1502 in order to have sole use of its benefits for restocking provisions. This island was subsequently 'discovered' again, independently, by other European voyagers. Not every island is necessarily precisely charted yet. For example, the island of Annobon (or Pagalu), a distant, offshore part of Equatorial Guinea is shown on British Admiralty chart 1595 published as recently as March 1995, with notes recording that the coastline and topography are derived from a map of 1913. Its position, though, is 'based on a ship's report of 1991', a quaint throwback to the earlier voyages of exploration. Chart 1595 also bears the rather chilling warning that 'mariners should exercise extreme caution when navigating in the vicinity'.

All existing islands might be known about now, but nature occasionally adds new islands to the list, usually through volcanic activity. Examples include Surtsey off Iceland in 1963; New Island in the inner bay of Deception Island in the South Shetlands which appeared in 1970 (Reynolds, 1995), an island off Iwo Jima in 1986 and a new volcanic island which emerged of the Solomon Islands in May 2000, captured on film by visiting geologists. Ferdinandea's peak is only eight metres below the Mediterranean and it could well emerge again if volcanic activity were to resume.

Although the existence of every island is known, it is not clear how many islands there are in the world. This seems odd, since surely anybody with access to a very comprehensive map repository, sufficient time and endless patience could simply count them. The problem is not one of location, however, but of definition, for it depends on what is counted as an island. At the large end of the scale, there is no problem. Australia at 7,686,843 km² is generally accepted as being a continent, which makes Greenland at 2,175,600 km² the world's largest island. The problem of definition comes at the other end of the scale. The basic dictionary definition of an island being a 'piece of land surrounded by water' (*Pocket Oxford Dictionary*) has the merit of being both simple to apply and comprehensive, but has limited utility in circumstances when some concept of scale or function is needed. A rock in the sea is a piece of land surrounded by water, but 'rock' has a separate definition in the dictionary as 'a mass of this [i.e. a solid part of the earth's crust] standing up into the air or water', so is a rock an island? And does an island have to be surrounded by sea at all times? Are periodic islands such as Lindisfarne off northeastern England or Mont St Michel off northwest France to be counted? The UK Hydrographic Office would say not, for to them only if the land is completely surrounded by water at high tide does it become an island.

Can an island graduate to become at least functionally an indistinguishable part of the mainland? Is Manhattan Island in New York in any real sense other than as an historical curiosity or through geographical pedantry still an island, given its myriad connections and complete integration with the surrounding mainland? Perhaps not, but nearby its less well-connected fellow City of New York Borough, Staten Island presumably is. Should Canada's province of Prince Edward Island have renamed itself Prince Edward Peninsular in 1997 when the Confederation Bridge opened to join it to New Brunswick?

The island of Mandø off the Jutland Peninsular of Denmark, which markets itself to tourists as being where *'der er meget mer'* (where the land meets the sea), is accessible by specially adapted road vehicles with large wheels that make it across the shallow coastal waters. Is Mandø thus an island? To the Vikings in the Dark Ages it presumably would not have been, for to them an island was only accepted as such if the passage between it and the mainland was navigable by a ship with its rudder in place. The Scottish census of 1861 defined an island as such only if it was inhabited and had sufficient pasturage to support at least one sheep (King, 1993). Thus an island had a value and a status; smaller places were of no account and were not regarded as islands. Often islands have had to be inhabited or at least landed upon to be regarded as having passed a functional or legal threshold to be granted some proper title. In sum, there is no sufficient definition of an island except in the 'land surrounded by water' sense and note that the dictionary also allows for 'isle' and 'islet' to be 'small islands', which further complicates matters. So our endlessly patient researcher in the map repository would end with a list of hundreds of thousands of pieces of land, all islands in the strict dictionary sense but many better regarded as isles, islets or rocks.

Discussion of what is an island and where islands are is not necessarily as arcane as it might seem; the Scottish idea of a functional definition makes that clear. Further, the International Convention on the Law of the Sea does not permit exclusive economic zones being awarded to rocks that cannot sustain habitation. Rockall is an outcrop of just 74 m^2 in the Atlantic, 461 km west of mainland western Scotland and 430 km north of northwest Ireland. To our Scots, it would not be an island, this bare crag having in itself no functional use. However, it was landed upon in 1955 by British marines lowered by helicopter to claim title for the UK, rather than remaining a rock unable to sustain habitation in international waters. Rockall was formally added to Scotland by *An Act to make provision for the incorporation of that part of Her Majesty's Dominions known as the island of Rockall into that part of the United Kingdom known as Scotland*, or the *Rockall Act of 1972*. In 1975 a former Special Forces soldier stayed on the rock for 40 days, further to establish British sovereignty. The contestation over Rockall relates not to the merits of the outcrop itself: if Rockall was a few hundred metres rather than a few hundred kilometres off the Scottish or Irish coast its presence would barely be remarked upon and nobody would need to land or stay on it. But if, 461 km from mainland Scotland, it is accepted internationally as a Scottish island, an integral, if very distant part of the UK, that country can claim rights to territorial waters and sea bed resources, and face down rival claims of ownership from Denmark, Iceland and Ireland. In 1974 the UK made just such a designation over 134,680 km^2 of continental shelf around Rockall. This area might well contain hydrocarbons – 30 oil companies are interested in the possibilities. The environmental group, Greenpeace, do not wish to see oil exploration extended to this sector of the Atlantic – and, to protest against any potential exploitation, themselves occupied Rockall in June 1997 with three men living in a high-tech, solar-powered, survival capsule. By

2 July 1997 Greenpeace had lived on Rockall longer than anybody else, and claimed sovereignty of what they declared to be 'Waveland'. Greenpeace stated that they did not want to own Rockall, but Deputy Executive Director, Chris Rose said:

> Four nations want the oil around Rockall. We do not recognise their right to develop it. We have told Tony Blair [the British Prime Minister] that we don't want Rockall itself but that the oil should be set aside for the common good. We are borrowing it until it is freed from the threat of development.
>
> (Greenpeace press release, 16 June 1997)

The Rockall case points up the difficulties of deciding what constitutes an island, as well as serving as one case of many in the world where islands/rocks are in disputed ownership, a theme that will be explored in Chapter 7. For current purposes it is sufficient to say that, whilst islands start at the large end with Greenland, there is no meaningful threshold at the small end. Take the example of French Polynesia. The local tourist board states that there are 115 islands in the territory's five archipelagoes, but that is counting each high island and atoll as a single island. However, atolls are broken up into different areas of land and if each separate 'area of land surrounded by water' were to be counted then French Polynesia has many thousands of islands. For example, tiny Tetiaroa atoll, one of the 115 islands, has, according to the same tourist board, 13 islets, although its accompanying map records 18 separate pieces of land. The 1:100,000 map of the Society Islands published by the Institute Geographique National has only 14 pieces of land for Tetiaroa. So does Tetiaroa count as one, 13, 14 or 18 islands? Also in the Society archipelago is mighty Tahiti, which at 1042 km^2 and reaching 2235 m is a 'high island' (i.e. with a mountain and fringing reef, as opposed to an atoll which has no mountain, just a reef (see Chapter 2)). There are also 13 *motu*, or dry areas on its fringing reef identified on the IGN map. Two of these, Motu Uta and Motu Tahiri have been incorporated by reclamation into the main island as part of the port at Papeete and part of the airport at Faaa respectively, so a case could be made for Tahiti to be one island or 14 islands or 12 islands. In functional terms it is Tahiti, one island; in geomorphological terms it may not be one island but is one system, the high island and its reef. To the west of Tahiti there is a single reef system enclosing two high islands, Tahaa and Taiatea: one system, two islands, or counting *motus* and detached parts of the two high islands, 47 islands. Logic says Tahiti is one island not 14; that Tahaa and Taiatea are two islands, not 47. However, were these places atolls – and in geological time this will happen to these high islands as it has done to others – then their *motus*, discounted now, would have to be regarded as islands in their own right. Those around former Tahaa and Taiatea would form just one atoll, one island or 40 plus islands. All that can be stated authoritatively is that we cannot state authoritatively how many islands there are.

Neither do we know how many islanders there are. Even if there were an agreed standard list of islands, it would still not be known how many inhabitants they had. This is because census statistics are by no means always presented in a way that identifies precisely those living on islands, especially small islands that often get collated into mainland administrative areas. Also small islands, especially, are often subject to periodic and seasonal swings in their population totals and thus their census population would reveal only one of a number of realities about their population size. Thus, it can only be estimated how many islanders there are in the world. The author once suggested that perhaps 9–10 per cent of the world's population live on islands, many in the four large insular nations of the UK, Japan, the Philippines and Indonesia (Royle, 1989). What cannot be challenged is that islands constitute an important part of the world's society and ecumene, as this book will demonstrate.

The concept of islands

In addition to their actual role in the world's economy and ecumene, islands have also played a part in human culture. The thought or concept of islands has informed literary, artistic, scientific and popular culture, as this chapter will now endeavour to show. This is the romantic notion of the 'island of dreams', the 'never-spelled-out popular mystique of islands' that Ramsey dates back to Classical times, before even what he identifies as the Mediaeval 'convention . . . that the site of anything marvellous was always an island' (1972, pp. 91–2). This romance will then be matched against realities throughout the book.

Two factors that make islands special are isolation and boundedness. Isolation is obvious – an island sits alone, any person wishing to visit must make a dedicated and unusual journey over water; they must leave the mainland, the familiar, and venture to the remote insular world. The attraction of this type of journey is what Baum (1997) calls the 'fact of difference'. He sees it in operation in visits across a political boundary as well and notes that as, say, within Europe, travel between different nations becomes easier and more familiar, the appeal of the fact of difference wanes. He then goes on to apply this idea to the island situation and opines that islands which build a fixed link (the term encompasses bridges, causeways and tunnels) lose the very 'islandness' which was part of the fact of difference, and, thus, their appeal. A journey on a road over a bridge is banal and has little sense of adventure. Baum cites a quotation about the debate on whether to build a bridge to Canada's Prince Edward Island:

> The idea of a fixed link joining Prince Edward Island to the mainland is a prospect that is disturbing to many islanders. The source of this disquiet is often so primal that it is not easily expressed, and it is sometimes articulated in terms of a threat to the 'Island way of life'.
>
> (Weale, 1991, p. 81)

Similar sentiments were expressed in Scotland during the same period when the Isle of Skye was joined to the mainland by a bridge. Much of the controversy about the Scottish development related, and still relates, to the high cost of the toll to use the bridge, but there was also a lobby against it being built simply because it took from Skye its islandness. This was particularly felt here, perhaps, because of the fame of this island with regard to the 1745 rebellion after the Scottish Jacobites' defeat at the Battle of Culloden in 1746 against the English. Their leader, the romantic hero figure (in legend anyway), Charles Edward Stuart – Bonnie Prince Charlie – escaped 'over the sea to Skye' in the immortal words of the *Skye Boat Song*. Going 'over the sea' on an unimaginatively designed, over-priced toll bridge does not quite convey the same image. Baum considers that 'an artificial land link removes the perfection of the island' (1997, p. 24). Prince Edward Island is then not perfect, for its Confederation Bridge to mainland Canada opened in 1997 (Figure 1.7). Peckham (2001) discusses the unpopularity of the plans of the late nineteenth and early twentieth centuries to 'unisland' England and that project was not carried out until many decades had passed. Off Ireland the largest island of Achill has had a causeway since 1886. This has, doubtless, been to the island's convenience, but Achill has lost its islandness; the mystery which in the nineteenth century saw the Protestant churches send missionaries there, as to 'darkest Africa', has gone and Achill is now just functionally a peninsular, like the Belmullet peninsular to its north. Meanwhile to its south, Clare Island, with its dramatic Knockmore mountain, stands an isolated sentinel across Clew Bay and ferry services now compete with one another to take tourists out to it.

Regarding the concept of boundedness, Weale has a quote from Lord Tweedsmuir, Governor General of Canada, in 1939, which usefully encapsulates the idea:

> What is it that gives an island its special charm . . .? I think the main reason is that an island has clear physical limits, and the mind is able to grasp it and make a picture of it as a whole.
>
> (Weale, 1992, p. 93, cited in Baum, 1997, p. 24)

Boundedness, one reason behind Baum's 'perfection of an island', combines with his 'fact of difference' to make islands special, to the appeal of writers and artists as well as inquisitive visitors. The physical consequences of insularity upon ecosystem development, as well as the curiosities of island formation, have made islands the subject of study of many scientists, including Darwin and Wallace, as will be discussed later in this chapter.

Islands in the artistic imagination

In *The Lake Isle of Innisfree* (1893) Irish author, W.B. Yeats fantasises about escaping civilisation to live alone on a small island. The ancient school primer in which the author has this poem, states that 'whether Innisfree is a real island or an imaginary refuge for the soul is of no consequence at all, so far as the spiritual significance of the poem is concerned' (The English Association, 1915). Maybe not, but to the Sligo Tourist Board it does and trips to Innisfree Island

Figure 1.7 Confederation Bridge, Prince Edward Island, Canada

in Lough Gill are available daily in the season from Sligo town. The island as a 'refuge for the soul' is a good encapsulation of the appeal of islands for those seeking succour of this nature. They can be poets, ordinary people, or hermits – other Irish islands, such as Skellig Michael and Caher, have served as the location for monastic communities. The lasting appeal of the island concept is demonstrated by the fact that the longest running programme on British radio is *Desert Island Discs*. Guests are asked how they would cope with life as a castaway on a desert island with only their choice of eight records, one book and a single luxury to sustain them.

The appeal of islands is both fed by and feeds upon the use of the concept of island in reality or metaphor by artists and writers. Many people will have heard of, if not actually read, such books as *Treasure Island* (Stevenson, 1883), *Robinson Crusoe* (Defoe, 1719) or its later adaption *Swiss family Robinson* (Wyss, 1812–13). Further, as this more humble author gazes out of his office window, he sees the same view of the edge of the 300 m Antrim Plateau on Belfast's northwestern boundary that Jonathan Swift saw in the early eighteenth century. To Swift, the undulations in the surface of the rounded mountains brought to mind a huge man lying on his side and thus the idea of Gulliver and his tiny tormenters in Lilliput was born. And where did Swift base his hero's adventures with both the tiny Lilliputians and the giant Brobdingnagians and the other less well remembered populations of his fantasy, *Gulliver's travels* (1726)? In mysterious, isolated islands where the standard biological processes that produced people of human size need not have applied. His island of Leputa was permitted even to traverse the laws of physics as well as biology, for it was a flying island.

Writers have found islands convenient places to allow characters to become free from normal social constraints and permit their more 'primitive', because less socially-constrained, sides to emerge. One of the best known examples of this is William Golding's *Lord of the flies* (1954). Here, a group of schoolboys are stranded on an uninhabited island. Initially, they struggle to reproduce their British society but in their isolated setting soon turn to barbarism, replete with bloody rituals and taboos. Stevenson's *Treasure Island* was remote, inaccessible, a good place to hide treasure away from prying eyes. Further, how better to study Crusoe's response to loneliness (no eight records for him) and isolation and his relationship with Man Friday when the latter appears, than for Defoe to trap him on the bounded ecumene of an island where the social mores of his past need no longer apply? Thus, most recent discussions of the novel depict Crusoe and Friday as 'relating homoerotically to each other before . . . [Crusoe returns] to white heterosexual civilisation' (Woods, 1996, p. 36). The author of *The Swiss family Robinson*, by contrast, uses the island adventure to make religious points, a feature of the book often omitted in modern abridgements. The fact that Robinson Crusoe was modelled on a real person adds romance and verisimilitude to the story. His alter ego was sailor Alexander Selkirk, who was marooned in 1804 for over four years on Másatierra, now sometimes called Robinson Crusoe Island, in Chile's Juan Fernandez Islands, after a dispute with his captain.

Some writers' accounts of islands have verisimilitude because of their own island upbringing. One example would be Liam O'Flaherty of Inishmore, one of the Aran Islands of Ireland's County Galway. O'Flaherty was the ninth child of an island farmer who supported his family from about six hectares of stony land. One obituary stated that O'Flaherty's novel, *Famine* (1937), was the best of his works (*Independent*, 10 September 1984), but *Skerrett* (1932), set on 'Nara' (anagram of Aran) from the 1880s has some telling remarks on island life which will feature in Chapter 4.

Other authors have used islands simply to create a manageable scenario for their plot and a natural limit to their cast of characters. Detective writer Agatha Christie used country houses to incarcerate her characters (*Ten little Indians* (originally *Ten little niggers* (sic) (1939)) or trains (*Murder on the Orient Express* (1934)); just as convenient were small islands, for example *Evil under the sun* (1941) set on 'Smugglers' Island'. Modern filmmakers have also used islands for similar purposes, thus in the year 2000 the BBC showed *Castaway*, produced by Lion Television. This is a 'docusoap' featuring 35 volunteers taken to live for a year on an abandoned island, Taransay in the Outer Hebrides.

Of greater note, perhaps, are more challenging works that used an island setting to create an ideal community according to the political theories of their authors. Thus, Sir Thomas More's model for the original *Utopia* (1516), was set on the isolated otherworldliness of an island. Shakespeare's play, *The Tempest* (1611) is famously set on an island. Its location

provided the setting for a bewildering variety of speculation about the
Edenic qualities of the island and the potential it offered for erecting an
alternative Utopian society on the one hand and for starkly encountering
the difficulties of sheer animal survival on the other.

(Grove, 1995, p. 34)

Other authors who have used islands to create alternative societies are H.G.
Wells (*The island of Dr Moreau* (1896)); and J.M. Barrie (*The admirable Crichton* (1909)).

An island is not the only literary setting that can allow an author to escape
from social, biological or even evolutionary norms, of course. Some writers have
used alternative devices to remove their characters from mainstream society to
make their political points, such as the future (e.g. Aldous Huxley's *Brave new
world* (1932)) or fantasy. Of the innumerable fantasy novels, perhaps the most
notable is George Orwell's *Animal farm* (1945), where the animals take on
human characteristics in a way that satirises the development of Stalinism.
Orwell (real name E.A. Blair), like Huxley, also used the future for another of
his political novels, *Nineteen eighty-four* (1949). Perhaps significantly, though
not set on an island, this was written on the otherworldliness of a remote island,
Jura, in the Inner Hebrides of Scotland, an example of a writer seeking 'refuge
for the soul'. Contemporary novelist William Sutcliffe, suffering from writer's
block, sought refuge in the Greek Cyclades: 'my main aim was to find a peaceful
island where I could relax and think things through'. This turned out to be
Amorgos and his block was cured, very directly; Amorgos becoming a location
used in the final chapter of the novel he was writing, *The love hexagon* (2000)
(*Independent*, 8 January 2000). Robert Louis Stevenson, of *Treasure Island*
fame, also went off to visit and live on islands: the Marquesas, the Paumotus,
the (then) Gilberts and, most notably, from 1890 until his death in 1894,
Samoa. Stevenson (e.g. 1896) wrote extensively about the Pacific. Painters, too,
made similar journeys for artistic and inspirational reasons, often to get back, as
they saw it, to a more primitive state. Paul Gauguin is the best example. He
travelled to various islands in Polynesia, especially the Marquesas. Belgian writer
Jacques Brel also spent the last years of his life on the Marquesas and he and
Gauguin are both buried in the Calvaire cemetery on Hiva Oa. There is a
museum of Gauguin's life and works on Tahiti.

Finally in this section, one might consider travellers' tales, such as, say,
Captain Cook's voyages in the South Seas. The modern versions of such travellers' tales are television programmes or films. Few will not have seen a screen
version of the 1789 mutiny of men from *HMS Bounty* led by Fletcher Christian
against Captain Bligh. Christian set Bligh and the loyal crew members adrift in
an open boat. The mutineers took the *Bounty* to enjoy the wonders of Tahiti
(for which the neighbouring island of Moorea stood in some filmed versions of
the story) before some of them incredibly went on to find their own private
South Sea hideaway in Pitcairn Island. The book of the story (Nordhoff and
Hall, 1932) was filmed in 1935, 1962 and 1984. The earliest and, to many

people's view, best version matched Charles Laughton as Bligh against Clark Gable's Christian. (See Loxley, 1990, on the literature of islands.)

Islands of dreams

The island as paradise concept is burnt into the psyche of the denizens of the western world at least, whilst they enjoy their pampered lives, mostly in the safety and security of the continental heartlands. 'Bounty' is the name of a British chocolate-coated coconut bar, the name and its advertising redolent of the romance of the South Sea islands, its 2000 campaign featuring a picture book entitled 'Atoll' which comes to life. Actually on Pitcairn (which is not an atoll) the nine mutineers of the *Bounty* crew, six Polynesian men, their 12 Tahitian women and then one baby found life so stressful on their tiny island from 1790 that they slaughtered each other. (Thus, they went just beyond the behaviour of the modern castaways on Taransay.) When Pitcairn was rediscovered in 1808 there were 20 or more children, but only nine women and one man, mutineer John Adams, were left alive. Two women had died in cliff falls, one from natural causes. Of the 14 male deaths, one was from natural causes, one from suicide, the other 12 were murders (Lummis, 1997). This reality is an unwelcome distortion of the romantic image. Instead the modern island

> castaway is confirmed as the master of all that he [sic] surveys. And if this is a fantasy, it is one that can be bought. Take a look at the travel brochures and try to find one that is not trying to sell you the desert island of your dreams.
>
> (Woods, 1996, p. 37)

To further illustrate this point, the author looked not at brochures but clippings from newspapers' travel sections about islands. The same vision of the island of dreams was found. The Pacific island scene generally is 'paradise' (*Independent*, 21 April 1986) including, presumably, the separate paradises of New Caledonia (*New York Times*, 17 November 1985) and Honolulu (*The Times*, 18 January 1986). By contrast, others have firmly located 'paradise' in the Caribbean on Guadeloupe (*The Times*, 3 December 1983). The wealthy, of course, have their own reserved piece of this Guadeloupe paradise on St Barthélémy (St Barts). This island is a 'tiny paradise for the chic' (*Independent*, 6 May 1995) and is sited at some physical as well as social distance from the island of Guadeloupe itself, where paradise was available at lower prices for the non-chic. As Guadeloupe has possession of paradise, you cannot actually reach it on St Lucia, though you 'won't get closer than this' to it there (*Independent*, 19 June 1993). You can also go 'lotus eating in the Grenadines' (islands shared between St Vincent and Grenada) (*Independent*, 17 March 1990). Even further from Guadeloupe is Grenada itself, not paradise at all in the author's collection of clippings, but still very pleasant, in fact with a wonderful inversion of hydrological reality, this island is 'an oasis in the sun' (*Belfast Telegraph*, 1 November 1997). So paradise is not in the Indian Ocean, but Nosy Bé off Madagascar is 'stranger than paradise' (*Independent*, 5 December 1992). The Seychelles possess not the real thing either for, with a nice mixture of island references, these islands are only 'a bounty bar paradise', (*Independent*, 30 December 1997). Rather, the Indian Ocean is the location of 'Eden', this being on Praslin in the Seychelles (*Independent*, 1 October 1989). On either side of that ocean are Phuket Island (Thailand) and Lamu Island (Kenya) both of which are 'idyllic' (*Independent*, 17 September 1994 and 21 November 1987, respectively). South of Phuket, Langkawi, just across the Malaysian sea border, is just 'mystical' (*Independent*, 24 August 1997).

The quotations identify the habit of journalists to slip readily into the use of tired clichés, and one would not claim these descriptions are profound analyses of the environment and lifestyles of the islands listed. To be fair to the *New York Times*, it should be mentioned that New Caledonia's 'paradise' was 'divided' and 'deadly', the article being written at a time of great tension and some violence between different ethnic groups on this French possession. However, it cannot be gainsaid that there is constant use of glowing terms to describe islands, and this, however remote from actual reality, is significant as it informs the public's perception about such places: islands are paradises; islands are romantic. And in a world where the biggest industry is tourism and where, as Chapter 9 will demonstrate, tourism has become many islands' staple economy, such perceptions are priceless. This author owns a T-shirt proclaiming him to be a 'friend of Rathlin', an island off Northern Ireland. Even friends of Rathlin must admit that this island is a workaday place, truly no paradise. However, such is the romance and mystique of islands that the local newspaper was able to record in some detail the wedding of a couple from the Irish Republic who were attracted to hold their ceremony on Rathlin on the 'island of dreams' principle (*Belfast Telegraph*, 27 October 1997). In a similar vein, couples can take advantage of the services of 'The Love Company' (ring 800–442–LOVE), which will arrange for their ceremonies (vows on request or create your own) to be held on the 'island of romance', also known, more prosaically, as Santa Catalina Island, California. In the South Pacific, the French Polynesian island of Bora Bora is also heavily marketed as a honeymoon destination and has the advantage of being considerably more attractive than either Santa Catalina or Rathlin.

Even more chic than to rent a piece of paradise on St Barts is to own your own. Many islands throughout the world are of a scale that they can be bought and sold in the same way as farms or estates. For example, the sixth Earl of Granville owns a 24,000 ha estate on Unst in the Shetland Islands. Because of the traditions of estate ownership in Scotland, many of the islands there can be bought and sold over the heads of resident tenants. In recent years sales have taken place of Dun Moraig, Easdale, Eigg, Eilean Aigas (formerly owned by popular historian Lady Antonia Fraser), Eilean Righ (formerly owned by Sir Reginald Johnston, sometime tutor to Pu Yi, the last emperor of China), Eilean Shona, Eriskay, Gigha, Grunay (whose 23.5 ha include a nine-hole golf course), Holm of Huip, Holy Island, Killegray, Langay, Little Bernera (possession of

Count Merrilees), Mingay and Clett (both of which once belonged to 1960s pop star, Donovan), Rum, Staffa, Tanera Mhor and Vacsay.

New owners are not always traditional Scottish lairds. A Punjabi prince bought Vacsay to use for meditation. An American businessman bought Staffa, site of Fingal's Cave, inspiration for Mendelssohn's *Overture to the Hebrides* of 1829, for his wife's 60th birthday. To be fair it must be recorded that he handed it over to the National Trust for Scotland afterwards with the proviso that his wife be named steward for her lifetime.

Some new owners bring changes to their islands. George Bullough, an industrialist from northwest England, owned Rum at the end of the nineteenth century and built the grand and dramatic Kinloch Castle that still dominates the island. Not content with changing the landscape, environment, economy and society, he even changed the island's name to Rhum; Rum, without an 'h', being insufficiently genteel for his *nouveau riche* sensibilities.

Not far away from Rum is Eigg, whose traditional owners sold it at the end of the nineteenth century to pay gambling debts. Eigg had stability from 1925 to 1966 when Sir Walter Runciman owned it, although the island was in slow decline and was described in 1972 as being silent, 'the silence of an introspective and depressed people, of a falling population, of lack of possibility' (Crossley-Holland, 1972, pp. 201–2). Owners after Runciman have included a Liechtenstein-descended English businessman and a German mystic artist. The businessman Keith Schellenberg, a former bob-sled champion, bought the island in 1975 and used it as a 'playground ... pootling round in his 1927 Rolls-Royce Phantom with screaming yahs waving champagne bottles out the window' (Johnson, 1992, p. 69). Schellenberg tried to make the island self-sufficient, but disputes with residents over security of tenure, the sale of some property and the provision of amenities made him unpopular and, one night in 1994, the Rolls-Royce was burnt out. Schellenberg started court actions against two newspapers, which had criticised his stewardship of Eigg but abandoned his suit in 1999, losing about £750,000 (*Independent*, 20 May 1999). Schellenberg sold Eigg for £1.5 m to the artist, Marlin Maruma. Maruma visited the island only twice before putting it on the market again (through Vladi Private Islands, see below) 15 months afterwards in July 1996, having sold off most of the island's cattle and having apparently reneged on promises to invest £15 m in the island. The asking price was then £2 m and the tenants, operating as a trust, stopped being silent and managed to find the money after a massive public appeal.

Also departing from traditional ownership, but far from being a playground, Scotland's, Holy Island was bought by Tibetan Buddhists in 1992 to become a centre for pilgrimage, 'continuing in the spiritual tradition' (Samye Ling Tibetan Centre, n.d.) of the life of island hermit, St Molaise.

Life on a tiny island can now be made convenient with solar energy, seawater desalination and prefabricated housing. The only problem is providing the money for the infrastructure – and for the real estate, for the demand for private islands is high and the prices, therefore, steep. The Buddhists of Holy Island had to raise £280,000 for their uninhabited place of pilgrimage. Eilean Shona in Argyllshire, which is 3.6 km long and has pine groves, an oyster farm and a nine-bedroom house, was on offer at £1.5 m in 1993. Tiny Holm of Huip off Orkney at only 25 ha and without buildings was priced at £100,000 in 1993.

Private islands may become paradises for the super-rich. Thus the man who developed the Virgin commercial empire, Sir Richard Branson, owns Necker Island in the British Virgin Islands (though the simply very rich can arrange to rent it). Former tennis player Bjorn Borg owns Kattilo near Stockholm; actor Marlon Brando, Tetiaroa in French Polynesia (first noticed by him whilst filming his version of *Mutiny on the Bounty*) though day-trippers from Tahiti can access it. The reclusive Scottish businessmen, brothers David and Frederick Barclay, own Brecqhou in the Channel Islands, which they bought in 1993 for £2.3 m and on which they built a £27 m castle. Actor John Wayne used to own Taborcillo off Panama, up for sale in 1998 for £195,000. Baron Rothschild, whose family name is a synonym for wealth, owns Bell Island in the Bahamas. Lambay Island near Dublin in Ireland, having been acquired by his parents in 1904 who had Sir Edwin Lutyens remodel the house, was the (very) private home of the fourth Baron Revelstoke from 1934 until his death in 1994 and he is buried there.

Vladi Private Islands, based in Canada and Germany, sell about 40 islands a year and normally has around 300 islands on their books. The most expensive sold by 1998 was for £12 m in the British Virgin Islands. The owner, Farhad Vladi, has four islands himself, and puts their popularity down to the fact that owners 'control what [they] see' (Mulholland, 1998, p. 60). Privacy is very important to some island owners; it was to Baron Revelstoke, for example, and is, also, to the Barclay brothers who in 1996 took a journalist to court for making an unauthorised visit to Brecqhou. Vladi's business strategy seems facile: 'islands look set to become the last unspoilt places on earth. Their popularity can only increase, which means prices will sky-rocket' (Hanssen, 1996, p. 44). The firm does not deal with islands that are likely to be subject to environmental, economic or political turmoil or where private ownership will offend locals, which rules out many in Asia and the tropics. Thus on the Seychelles there were in 1998 still claims outstanding by foreigners (including former Beatle George Harrison and arms dealer Adnan Khashoggi) whose land was appropriated in 1981 by the government (*The Times*, 1 January 1998). In cases where an island does meet the right criteria, is attractive to rich buyers seeking to get away from it all and wanting to become owners of all they survey in a controlled, private, secure environment, 'you are talking about a crown jewel' (Hannsen, 1996, p. 45). Coconut Island, off Oahu in Hawaii, was sold through Vladi Private Islands for £5.7 m – the price of paradise is high, indeed.

Islands and science

In addition to having a special place in the literary and cultural world, islands have also a distinct role in science, though one presumably free from romantic associations. In his important book on the early history of environmentalism, Richard Grove adds the significance of the island as a concept in the development of science, reinforcing the place he acknowledges it already had in the development of culture. His first chapter is entitled 'Edens, islands and early empires' (1995; see also Schulenburg (2001) who also uses the association of Eden, islands and culture). This, again, comes about largely through the operation of those insular qualities of isolation and boundedness. King puts it thus:

> For geographers, anthropologists, ecologists and biologists, islands hold a particular attraction, functioning as small-scale spatial laboratories where theories can be tested and processes observed in the setting of a semi-closed system.
>
> (1993, p. 13)

The small-scale idea was seen in the early major work on island science by Alfred Russell Wallace: 'in order to limit the field of our enquiry . . . we propose to consider only such phenomena as are presented by the islands of the globe' (Wallace, 1892, p. 241). The application of the island as laboratory can be seen most significantly in the work of Charles Darwin. Darwin's voyage on *H.M.S. Beagle* from 1831 to 1836 took him to a number of island groups, including the Falklands in 1833 and, more importantly, the Galapagos Islands off Ecuador in 1835. Here variation in environment was seen to be reflected in distinctions between related species. In particular, Darwin became fascinated by the varieties in beak form of a number of types of finch found on the different islands. The giant tortoises were also different on each island, with those on less favoured islands having an ability to reach higher, to be able to graze from trees rather than just ground vegetation. Observations on the Galapagos Islands helped Darwin to develop ideas on biological adaption and were significant also in his work on evolution, his theory of natural selection, published much later (Darwin, 1859). In addition to informing his thinking on evolution, islands also stimulated Darwin's work on geology and he produced two books containing material about the nature and formation of oceanic islands (1842; 1844). Wallace had meanwhile come independently to conclusions on evolution similar to Darwin's (Wallace, 1855; see also Grant, 1998) and the two men presented their findings together at a lecture at the Linnaean Society in London in March 1858. That the idea of the island as laboratory had also played a part in Wallace's thinking can be appreciated from his later book, *Island life*, in which he stated:

> Islands possess many advantages for the study of the laws and phenomena of distribution. As compared with continents they have a restricted area and definite boundaries, and in most cases their biological and geographical boundaries coincide. The number of species and of genera they contain is always much smaller than in the case of continents, and their peculiar species and groups are usually well defined and strictly limited in range. Again their relations with other lands are often direct and simple and even

when more complex are far easier to comprehend than those of continents; and they exhibit besides certain influences on the forms of life and certain peculiarities in their distribution which continents do not present, and whose study offers many points of interest.

(Wallace, 1892, pp. 241–2)

In the contemporary era, islands continue to be significant in one modern area of biology, namely genetics. Some island populations have been relatively isolated for a long time. This has given them collectively a restricted genetic range and this can be of considerable significance in medical research (see Chapter 4). Thus, the population of Iceland is being investigated to try to track down any genetic predisposition towards diseases. A long way south in the Atlantic, the people of Tristan da Cunha have a genetic predisposition towards asthma. An American company is trying to understand why in the hope of developing a better understanding of the nature of this increasingly significant disease.

Island realities

Generally speaking, are islands paradises or is this just hype, a mixture of people's desire for romance and mystery, boosted by popular culture and advertising? For one answer we can return to island literature, this time not to authors using islands as a convenient location for their plot development or on which to base their social theories, but to writers from islands discussing their own experiences. This can be appreciated most readily from the trilogy of auto-biographies of residents of Great Blasket Island off western Ireland published between 1929 and 1936 (O'Crohan, 1929; O'Sullivan, 1933; Sayers, 1936) which are examined in detail in Chapter 5. From their books it is clear that there was a lot to commend life on their island, particularly the richness of its society and oral culture as well as the companionship of its isolated population. However, economically and materially life was hard. All three books are much taken up with the constant emigration of islanders seeking a better life else-where and, indeed, the island was finally evacuated in 1953.

Finally, to be fair to the journalists whose laboured and cliché-ridden words of praise for islands were mocked above, one must report that, of course, news-papers carry, if less frequently, more realistic pictures of island life. Noted jour-nalist and commentator Philip Howard wrote that:

Life on an island may seem romantic to susceptible townies; in reality it is narrow, cabin'd, cribb'd, confin'd, bound in. Your shepherd who does it for a living takes a far more robust view of the supposed pleasure of trudg-ing up the mountain again than your wet romantic. Tenants of Ailsa Craig used to pay their rent in gannet feathers. Have you ever tried to part a gannet from its feathers? I should care to have to do so for the rent.

(*The Times*, 1 August 1984)

Further, at the end of an article about a young couple who moved from London to open a guesthouse on St Agnes, the most isolated of the Scilly Isles off southwest England, appeared this paragraph:

> As the boat pulls away from the quay, leaving Rachel and Piers waving frantically in their oilskins, my thoughts are already turning to London. I am not sure if I have left them in paradise or in prison.
>
> (*The Times*, 1 January 1998)

The answer could be somewhere in between, or it may be both. An island that might be a paradise to, say, a person on vacation, may at the same time be a prison to an islander seeking opportunities or services not available there. The other point to be taken from the paragraph is the fact that the journalist's thoughts were returning to London. Islands are quintessentially peripheral and thus their people are irredeemably provincial. This view is not confined to London journalists, or to the modern era. Thus in a seventh-century dispute over religious observances, to Cummian, an authority based in Rome, the Columban community on Iona in Scotland's Inner Hebrides were described thus:

> An insignificant group of Britons and Irish who are almost at the end of the earth, and if I may say so, [are] but pimples on the face of the earth.
>
> (Cummian, c. 632, cited in Walsh and Ó Cróinín, 1988, p. 75)

Iona has remained a religious place, the monastery dating back to St Columba has been restored, and there is a religious community living there to this day.

Islands may not necessarily always be paradises but what is not in dispute is 'the geographical fascination of islands', to make use of the title of Russell King's essay on the subject and the last words of this chapter can be left to him:

> An island is a most enticing form of land. Symbol of the eternal contest between land and water, islands are detached, self-contained entities whose boundaries are obvious; all other land divisions are more or less arbitrary. For those of artistic or poetic inclination, islands suggest mystery and adventure; they inspire and exalt. On an island, material values lose their despotic influence; one comes more directly into touch with the elemental – water, land, fire, vegetation, and wildlife. Although each island naturally has its own personality, the unity of islands undoubtedly wields an influence over the character of the people who live upon them; life there promotes self-reliance, contentment, a sense of human scale.
>
> (1993, p. 13)

References

Key readings

For the significance of islands in scientific development, see Darwin and Wallace, also Grove and Grant for more recent discussion and commentary. Regarding island literature, the most significant is Defoe's *Robinson Crusoe*. For a realistic view of life by an islander, any of the three Great Blasket autobiographies should be consulted, the author's favourite is that of O'Sullivan. See Loxley for a commentary on island literature. For a range of views on island development, consult Lockhart, Drakakis-Smith and Schembri.

Baum, T. (1997) 'The fascination of islands: a tourist perspective', in D.G. Lockhart and D. Drakakis-Smith (eds) *Island tourism: trends and prospects*, Pinter: London, pp. 21–36.

Barrie, J.M. (1909) *The admirable Crichton*, Upcott Gill: London.

Christie, A. (1934) *Murder on the Orient Express*, Collins: London.

Christie, A. (1939) *Ten little niggers* (sic), Collins: London.

Christie, A. (1941) *Evil under the sun*, Collins: London.

Crossley-Holland, K. (1972) *Pieces of land: journeys to eight islands*, Victor Gollancz: London.

Darwin, C. (1842) *The structure and distribution of coral reefs, being the first part of the geology of the voyage of the Beagle*, Smith, Elder: London.

Darwin, C. (1844) *Geological observations on the volcanic islands visited during the voyage of the Beagle*, Smith, Elder: London.

Darwin, C. (1859) *On the origin of species by means of natural selection, or the preservation of favoured races in the struggle for life*, John Murray: London.

Defoe, D. (1719) *The life and strange surprising adventures of Robinson Crusoe of York, mariner*, modern edition (1983) Oxford University Press: Oxford.

English Association (1915) *Poems of today*, Sidgwick and Jackson: London.

Golding, W. (1954) *Lord of the flies*, Faber: London.

Grant, P. (ed.) (1998) *Evolution on islands*, Oxford University Press: Oxford.

Grove, R.H. (1995) *Green imperialism: colonial expansion, tropical island Edens and the origins of imperialism, 1600–1860*, Cambridge University Press: Cambridge.

Gutsche, A. and Bisaillon, C. (1999) *Mysterious islands: forgotten tales of the Great Lakes*, Lynx Images: Toronto.

Hanssen, H. (1996) 'Paradise for sale', *Geographical Magazine*, 68, 10, pp. 44–5.

Huxley, A. (1932) *Brave new world*, Chatto and Windus: London.

Johnson, R. (1992) 'Eigg's for sale', *Sunday Times Magazine*, 1992, pp. 63–70.

King, R. (1993) 'The geographical fascination of islands', in D.G. Lockhart, D. Drakakis-Smith and J. Schembri (eds) *The development process in small island states*, Routledge: London, pp. 13–37.

Loxley, D. (1990) *Problematic shores: the literature of islands*, St Martin's Press: New York.

Lummis, T. (1997) *Pitcairn Island: life and death in Eden*, Ashgate: Aldershot.

More, T. (1516) *Libellus vere aureus nec minus salutaris quam festiuus de optimo reip. statu, deq[ue] noua insula Vtopia*, T. Martin: Lovain (first translated into English in 1551; modern edition, More, T. (1996) *Utopia*, Phoenix: London).

Mulholland, R. (1998) 'Island shopping', *Independent Magazine*, 1 November, pp. 60–3.

Nordhoff, C. and Hall, J.N. (1932) *Mutiny on the Bounty*, modern edition (1983), Corgi: London.

O'Crohan, T. (1978) *The islandman*, Oxford University Press: Oxford. (First published as Ó Criomhthain, T. (1929) *An tOileánach*, Oifig an tSoláthair: Baile Átha Cliath.)

O'Flaherty, L. (1932) *Skerrett*, Wolfhound Press: Dublin (1982 reprint).

O'Flaherty, L. (1937) *Famine*, Wolfhound Press: Dublin (1979 reprint).

Orwell, G. (1945) *Animal farm*, Secker and Warburg: London.

Orwell, G. (1949) *Nineteen eighty-four*, Secker and Warburg: London.

O'Sullivan, M. (1953) *Twenty years a'growing*, Oxford University Press: London. (First published as Ó Súilleabháin, M. (1933) *Fiche blian ag fás*, Clólucht an Talbóidigh: Baile Átha Cliath.)

Peckham, R.S. (2001) 'The uncertain state of islands: nationalism and the discourse of islands', *Journal of Historical Geography*, forthcoming.

Poe, E.A. (1838) *The narrative of Arthur Gordon Pym of Nantucket*, Harper and Bros: New York.

Ramsay, R.H. (1972) *No longer on the map: discovering places that were never there*, Ballantine Books: New York.

Reynolds, J. (1995) 'Land of ice and fire', *Geographical Magazine*, 67, 11, pp. 38–9.

Royle, S.A. (1989) 'A human geography of islands', *Geography*, 74, pp. 106–16.

Samye Ling Tibetan Centre (n.d.) *Help us buy Holy Island as a place of prayer and retreat*, Samye Ling Tibetan Centre: Eskdalemuir.

Sayers, P. (1974) *Peig: the autobiography of Peig Sayers of the Great Blasket Island*, Talbot Press: Dublin. (First published as Sayers, P. (1936) *Peig*, Clólucht an Talbóidigh: Baile Átha Cliath.)

Schulenburg, A.H. (2001) ' "Island of the blessed": Eden, Arcadia and the picturesque in the textualising of St Helena', *Journal of Historical Geography*, forthcoming.

Stevenson, R.L. (1883) *Treasure Island*, Cassell: London.

Stevenson, R.L. (1896) *In the South Seas*, Chatto and Windus: London.

Sutcliffe, W. (2000) *The love hexagon*, Hamish Hamilton: London.

Swift, J. (1726) *Gulliver's travels*, B. Motte: London.

Wallace, A.R. (1855) 'On the law which has regulated the introduction of new species', *Annals and Magazine of Natural History*, 16.

Wallace, A.R. (1892) *Island life: or the phenomena and causes of insular faunas and floras including a revision and attempted solution of the problem of geological climates*, Macmillan: London (first published in 1880).

Walsh, D. and Ó Cróinín, D. (eds) (1988) 'Cummian's letter: *De controversia Paschali*', *Pontifical Institute of Medieval Studies, Studies and Texts*, 86.

Weale, D. (1991) 'Islandness', *Island Journal*, 8, pp. 81–2.

Weale, D. (1992) *Them times*, Institute of Island Studies: Charlottetown.

Wells, H.G. (1896) *The island of Dr Moreau*, Heinmann: London.

Woods, G. (1996) 'Desert island desires', *Geographical Magazine*, LXVIII, 1, pp. 36–7. (Another version of this article appears in D. Bell and G. Valentine (eds) *Mapping desire: geographies of sexualities*, Routledge: London.)

Wyss, J. (1812–13) *The Swiss family Robinson* (first published in English (1852)), Simpkin, Marshall: London. (Originally published in German as *Schweizerische Robinson*.)

2 Islands
Their formation and nature

The distribution and formation of islands

Islands are distributed across the globe, but this distribution is far from even. Nor can the distribution be described as random, given that the appearance of any island is a logical response, indeed presumably an inevitable response, to a set of physical circumstances operating at that place at the time of formation and since. Island formation can basically be subdivided into two main causes. First there are islands that are appended to a continental fringe and whose origin and formation are tied in to changes at the regional scale, such as those operated through glaciation and all that is associated with it, including isostatic and eustatic adjustments. Continental fringe islands can also result from the local scale operation of forces, such as erosion, weathering and deposition.

Other islands, often further from the edge of the continents, result from the operation of global forces such as plate tectonics and its associated vulcanicity. In Europe, the volcano of Mount Vesuvius may be on the Italian mainland, but there are also volcanic islands in the region, for example Mount Etna on Sicily and, in the Isole Lipari or Aeolian Islands north of Sicily, Stromboli and the eponymous Vulcano Island itself.

Volcanic arc islands

As the earth's slowly moving tectonic plates meet, tremendous energy is unleashed. If the plates are grinding past each other in what is known as a transform fault boundary – as in southern California, where the Juan de Fuca plate is moving past the North American plate – these forces can result in earthquakes. If the plates meet head on at a convergent boundary, orogeny results, associated with volcanoes as well as earthquakes. There are three scenarios. If two continental plates meet, the relatively light and buoyant continental rocks buckle, as when the once separated landmass of India rammed into Asia and formed the Himalayas. There is no island formation there, nor are islands formed when an oceanic plate meets a continental plate and subducts under it. In this scenario huge linear mountain ranges form, as with the Andes where the lithosphere of the Nazca plate slides beneath the continental crust and lithosphere of the

South American plate. Vulcanicity also results as the material from the sub-
ducted plate melts and, as magma, rises and may find its way to the surface as
volcanoes.

Where two oceanic plates meet, islands are formed (Figure 2.1). One plate
subducts under another, as with the oceanic–continental collisions, and again
vulcanicity results. In this scenario the volcanoes are not part of a mountain
range, but have their roots on the ocean floor and, dependent upon sea level,
such volcanoes can break the surface, forming a volcanic island arc. Some way
off from this arc a deep oceanic trench will be found where the actual subduc-
tion is occurring (see, for example, Tarbuck and Lutgens, 1987; Wicander and
Monroe, 1995).

Thus, for example, the Japanese islands are to the west of the Japan Trench:
Mount Fuji, the great symbol of that nation, with its classic cone formation, is
an island arc volcano. The Alutian Islands rise to the north of the Alutian
Trench; the Marianas to the west of the Mariana Trench, all part of the Pacific
Ring of Fire. The curve of the Indonesian archipelago arises from a 7000 m
deep trench, the Sunda or Java Trench and the Philippines are to the west of
the Philippines Trench. The huge eruption of Mount Pinatubo on Luzon
Island in the Philippines in 1991, which killed 343 people and rendered
200,000 homeless, has been a recent example of the volcanic forces in action.
Not all the land in these islands necessarily has to be volcanic. Over geological
time the islands are eroded, their sediments added to the sea floor, perhaps even
reaching the ocean trench where metamorphism can take place. Long-standing
or mature island arcs such as Japan can and do contain a complex of volcanic
rocks, folded and/or metamorphosed sedimentary rocks and intrusive igneous

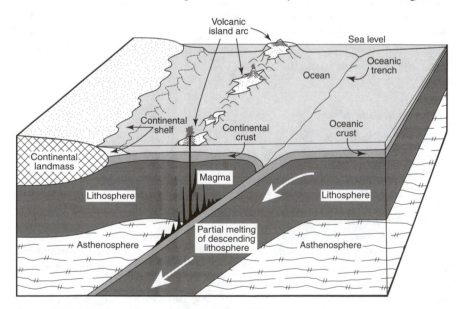

Figure 2.1 Island formation as oceanic plates collide

rocks. The forces of plate tectonics can also give rise to some of the isolated oceanic islands as will be demonstrated below.

Oceanic islands

Oceanic islands are those which arise from the ocean depths. Their origin is tied up with vulcanicity and plate tectonics, even though many oceanic islands, especially in the Pacific, are not themselves now constituted from volcanic material.

Some oceanic islands are 'fragments discarded from ancient continents' to use Mitchell's words (1989, p. 20). These include New Caledonia and New Zealand. Others are those formed by the upwelling of magma in the mid-oceanic ridges (Cann, Elderfield and Laughton, 1999). These upwellings form constructive tectonic plate margins. One classic example is the mid-Atlantic ridge which separates European and African plates from those of the Americas. From Jan Mayen Island, north of Iceland, the ridge runs southwest through the Azores, before swinging east around the Cape Verde Basin, reaching the surface at St Paul's Rocks in the north Atlantic. South of the equator it runs almost north–south via Ascension Island, Tristan da Cunha and the other islands in its group, before turning to the east and making its last appearance above sea level at Bouvet Island. These islands are all geologically recent products of the ridge. On Ascension Island's 97 km^2 there are 44 volcanic cones, mainly of small size (Figure 2.2), but including the mighty Green Mountain at 859 m. This mountain is the oldest part of the island and is high enough to be affected by orographic rainfall and receives over 500 mm of precipitation per annum. This has helped to break down the lava here and Green Mountain has soil; indeed it has some lush, tropical vegetation. Much of Ascension's landmass is lower, younger and dryer, with Georgetown on the west coast receiving a variable precipitation of between 25 and 178 mm annually (Oldfield, 1987). The lower part of the island is composed of smaller cones, recent lava flows and cinders. The surface is porous and soil formation has not progressed very far. Much of the area is not vegetated, but this situation is changing. As recently as 1987 Oldfield wrote of Ascension: 'Today, low altitude areas are barren with occasional grass and the endemic *Euphorbia origanoides*' (p. 80). A few years later the Ascension landscape had become much more verdant. Most of the new greenery is provided by a recently introduced exotic, Mexican thorn (*Prosopis juliflora*), the seeds of which are ingested by feral donkeys, whose wanderings carry them to new areas and provide them with a fertilised medium in which to sprout. Such pioneering spread of the first vegetation to become established over much of the island is a mark of how young this landscape is, as well as how isolated.

There has been no actual volcanic activity in Ascension's recorded history since its discovery in 1501, but in geological terms the landscape is recent and its vulcanicity cannot yet be regarded as extinct. Elsewhere on the ridge there is more activity. The volcano that is Tristan da Cunha last erupted in 1961, leading to the temporary evacuation of the island. To the north, Iceland abounds with hot springs and even geysers and has had very recent activity in

Figure 2.2 Volcanic features on Ascension Island, 1999

the frequent eruptions of Heckla on the main island. Outpourings of lava onto Heimaey in the offshore Westman Islands in 1973 led to 5000 inhabitants being evacuated for nearly six months. This was the eruption where seawater hoses were used to cool the viscous advancing lava, which saved the town centre and harbour. The harbour never closed and one fish-processing factory stayed working during the crisis. Off the Westman Islands, Surtsey Island arose from the sea in 1963.

Mid-oceanic ridges in the Indian Ocean have not led to the formation of many islands, except for the Mascarene Ridge which surfaces in high volcanic islands in the Seychelles, Mauritius, also Réunion, where the Piton de la Fournaise displays much evidence of recent activity. Further south the inhospitable Kerguelen, Heard and Macdonald Islands sit astride the Kerguelen Ridge. In the Pacific Ocean and surrounding seas there are relatively few mid-oceanic ridge islands. Easter Island (Rapa Nui) is on the East Pacific Ridge where the Pacific plate diverges from the Nazca plate.

Other oceanic islands, especially in the Pacific, owe their formation to hot spot vulcanicity (Figure 2.3). The theory here is that the earth's crust contains certain areas of weakness, which, at irregular intervals, will give way under pressure from the molten core beneath. The upper mantle and oceanic crust can be broken through and, at such times, there is an outpouring of lava that, dependent upon sea level again as well as upon the magnitude of the eruptions, can break the sea surface and form an oceanic volcanic island. The irresistible movement of the tectonic plate takes the young island on the lithosphere away from

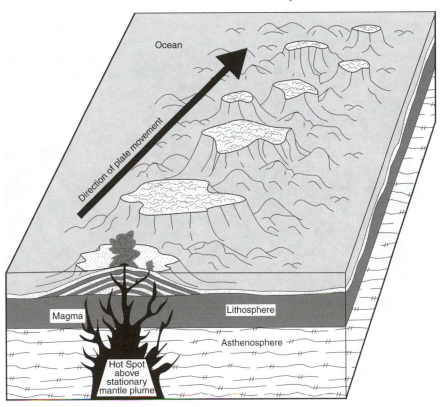

Figure 2.3 Island formation from hot spot vulcanicity

the hot spot. When the next eruption takes place, another island can form or
the original island be added to with the new material. Which happens depends
on the time between eruptions and their magnitude. As the process repeats itself
a chain of volcanic islands is formed, representing past and present, (if any)
activity over the hot spot. The Hawaiian Island chain is the classic case. Here
there are a number of volcanoes of different age on the big island, Hawaii itself.
The most recent and most easterly is Kilauea (Figure 2.4), actively erupting as
this is written, with molten lava pouring from its flank through a lava tube into
the ocean, with a permanent cloud of steam and noxious gases being produced
as a result. The other volcanoes on the island of Hawaii are to its northwest.
Continuing along that direction, for that is the line taken by the Pacific Plate,
are the other islands of the Hawaii chain such as Maui, Molokai and Oahu, all
removed from the hot spot. Young volcanic islands of this nature are formed of
light basaltic material, which has little resistance to erosional forces. Rainstorms
on Tahiti, for example, see the island's numerous waterfalls turn into torrents of
brown water tumbling down the precipitate shoreline into the sea. These high
islands are also subject to compaction and subsidence under their own weight,
typically declining at about one centimetre per century. Over time such islands

Figure 2.4 Kilauea volcano, Hawaii Island, Hawaii, USA, 1999

can slip into the sea and the Hawaiian chain dies out with a series of atolls and reefs such as Pearl and Hermes Atoll and Kure Atoll. Next come the Milwaukee Seamounts, and beyond them, the Emperor Seamounts (Wicander and Monroe, 1995). On rare occasions the mountain might rise again, to form a flat-topped *makatea* island (Mitchell, 1989).

In tropical seas these collapsing volcanic islands are associated with the creation of atolls. Charles Darwin was an early writer on their formation. Coral reefs will form in the right conditions of warmth, light and nutrients, both off continents, as with the Great Barrier Reef off Australia, and around high islands. Tahiti has a fine reef, with some of it under sea level but with areas rising above sea level thanks to the deposit and capture of broken coral. In Polynesia such reef islands are called *motu*. If there is relative sea level rise, as there would be with an island sinking, it is possible for the reefs to continue to survive. The living part of the reef can build upon the dead and dying part below; the sinking of the island taking the rooted reef down with it to depths where the coral animals cannot survive. Over time, the island sinking might see just the tips of the landmass within a now distant reef. In the end the island totally disappears and only the reef is left, and the *motu* islands and the reef form an atoll. This surrounds incompletely a lagoon with relatively shallow water over the sunken mountain. The water on the ocean side of the atoll/reef is much deeper. French Polynesia's Society Islands form a perfect example of this process. Tahiti and its neighbour, Moorea, are high islands with a surrounding reef representing fairly young volcanic activity. Tahiti itself is a double volcanic island with Tahiti Itu being another roughly circular mass that abuts onto the bigger, near-circular landmass that is Tahiti. To the west the islands are older and have begun to sink; Bora Bora and Maupiti have just the tips of the once extensive high islands' mountainous land mass forming islands within their lagoons. To the west again are true atolls, Maupihaa, Manuae and Motu One, with nothing breaking the surface of their lagoons. This region of French Polynesia thus has islands and atolls at all stages of this model. Many other Pacific nations are just atolls. Examples are Tuvalu, most of the separate island chains that make up Kiribati, as well as the Marshall Islands with its two rows of atolls, the Ratak and the Ralik chains.

Coastal islands

Current coastal islands have normally been created through the application of non-volcanic forces. They are often associated with the action of sea level rise following the ending of an ice age. Water, formerly locked into ice bodies such as glaciers on land or ice sheets at sea, melts and sea levels rise, with the result that some places once joined become separated. Thus there were land bridges between the European continent to what was to become the island of Great Britain and across to Ireland. There are current concerns about sea level rise consequent now not upon the ending of an ice age but upon global warming, probably resulting from mankind's activities. Rapid sea level rise could see the disappearance of low-lying island groups such as the Maldives, the Marshall Islands and many others. Sea level rise would also create new islands by the flooding of low-lying continental regions, which would inevitably surround some areas of relatively higher ground by water. This is compensation only in numerical terms; there is little that can be said in favour of sea level rise.

Many other coastal islands are clearly the product of erosion, the power of the sea or other forces separating them from their landmass. The sequence leading to the formation of sea stacks via caves and natural arches is familiar to students of geography from an early age. The inevitability of such progression can still come as a surprise, however, and small islands are obviously at particular risk from erosion forces, being by definition largely coastal zones. In the western part of Prince Edward Island in Canada, the economy is in some difficulty and the local development corporation had introduced a tourism marketing strategy based to an extent around 'Elephant Rock'. This was a large stack with a natural arch at its seaward end that made it look, indeed, like an elephant, the seaward arm of the arch being its trunk. Sadly, the coast here is of fairly soft sandstone and is subject to quite severe and rapid erosion. Over the Christmas holidays at the end of 1998 the elephant lost its trunk and the tourism strategy was devastated, although they made the best of it with Internet reports on the health of the beast and some people now go to see where the elephant had been. A suggestion that the trunk should be replaced by a fibre-glass prosthesis was rejected. In due course coastal erosion will inevitably produce another stack and another elephant might be born.

The coast of Guangxi province near Guilin in China is notable for its limestone pillar islands. These result from weathering and erosional processes in this karstic landscape, which have rendered deep beds of limestone into individual pillars and blocks. Where the base of the weathering activity is now below present sea level, islands result. Other limestone coastal islands are to be found elsewhere, such as off the coast of Croatia.

Glacial erosion can scour troughs into a landscape that, with subsequent sea level rise after the end of the ice age, become flooded, forming islands. One example is the eastern coast of Greenland, which is a magnificent, complicated tangle of islands and peninsulas. The ragged island fringe of southern Chile would also have been affected by glaciation in this way.

The Faroe Islands between Iceland and Scotland owe their origin to geology insofar as the overall landmass is the result of volcanic forces, basaltic outpourings during the tertiary era. However, the existence now and the actual separation of the Faroes into 18 principal islands, plus dozens of stacks and rocks, owe much to the geologically recent action of glaciation. This process cut troughs that have subsequently been flooded, forming impressive fjords and straits, which now separate the landmass into different islands. Stremoy Island, for example, is cut into by a number of fjords, but between Stremoy and Eysturoy and Stremoy and Vagar the troughs formed straits which now separate Stremoy from its neighbouring islands.

Glacial deposition can lead to the formation of islands, too. For example, in Ireland there are many thousands of drumlins, small rounded hills made up of deposits laid down under an ice sheet. In Clew Bay on the west coast a drumlin swarm has been flooded by sea level rise and this bay is studded with scores of small islands, which are the drowned drumlins poking their heads above the water. On the east coast the flooding of Strangford Lough has also led to a scatter of islands.

Depositional coastal processes not connected directly with glaciation can also form islands. Thus, the eastern coast of the United States has a fringe of islands, many of which are made up of coastal sand deposits. Deposition of riverine material can also form coastal islands when the energy of the river is dissipated by its arrival at the sea, upon which material carried in suspension or by traction is released and deposited. Under certain conditions deltas are formed and such deposits can and do see the formation of low-lying islands. Thus Egypt's Nile Delta has hundreds of small islands within it. Where the Ganges system hits the sea in the Bay of Bengal carrying huge loads of deposits down from the Himalayas, similar low-lying, silty islands form. Such is the pressure on land resources in Bangladesh that these islands are colonised by farmers and fisherfolk. Sadly they may only prove to be ephemeral features. The Bay is subject to severe flooding on a frequent basis through the operation of monsoonal weather systems and/or through excessive flows of water through the river system. On such occasions the coastal silt islands can be inundated or even washed away, with heavy loss of life a common result.

Depositional processes can actually take place some way offshore if the conditions, principally the interaction of currents, or other means whereby a stream of water is slowed, mean it is no longer able to continue to carry its load in suspension and it is deposited. Thus Sable Island is more than 150 km off Nova Scotia, but is entirely composed of sand, and its origin is the result of the forces of coastal erosion, transportation of sediment and deposition. Thus despite its distance from the landmass, it is still a continental rather than an oceanic island in terms of its origins. Such deposition can also remove insularity; tombolos are depositional landforms that connect what would have been a true island to a nearby coast or another island. St Ninian's Isle is a fine example from the Shetland group, whilst the Canadian Magdalen Islands/Isles de la Madeleine in the St Lawrence form an archipelago of what were once eight separate landmasses, all but one now linked by a complicated and extensive tombolo system.

The biogeography of islands

Islands are often rather unusual and interesting places in biological terms (see, for example, Chester and McGregor's *Australia's wild islands* (1997); also Quammen, 1996). Typically an island has a restricted range of species of both flora and fauna compared to those of a continental region with similar climatic and other characteristics. Islands are isolated; this means that some species may not have been able to reach the landmass. Great Britain has a more restricted fauna than France; Ireland, off Great Britain, has a more restricted fauna still, lacking snakes and moles, for example. The Irish offshore islands have a more restricted fauna than mainland Ireland. Similarly, Skomer Island off west Wales has only five species of mammals (excluding bats) compared to 39 on the mainland opposite. All were probably accidentally introduced by mankind.

Isolation can also cut species off and this may, given sufficient time, see them adapt to the specific challenges and opportunities of their environment by evolving into new forms. Thus, on Skomer Island, the introduced vole has become an endemic sub-species, the Skomer Vole:

> It is closely related to the bank vole found on the mainland with which it will interbreed. However, its larger size, lighter colour and differences in its skull and dentition, are sufficient for it to be considered an island race or sub-species. No doubt, if it remains isolated for a long enough period it will gradually evolve into a separate species.
> (West Wales Naturalists' Trust, n.d., p. 22)

On Socotra in the Gulf of Aden about one third of the 850 plants so far identified are unique. The Royal Botanical Gardens of Edinburgh were, in the mid-1990s, working on a blueprint to try and ensure that when, inevitably, increased development comes to the 40,000 islanders, minimal disruption will be caused to the ecosystem. Ascension Island is famous as a nesting place of the Green Turtle (*Chelonia mydas*) (Mortimer and Carr, 1984), but is significant to biology for other species, too. There are two endemic birds – the Ascension Frigate Bird (*Fregata aquila*) and the Red Footed Booby (*Sula sula*). There are ten endemic plants: *Anogramma ascensionis* (possibly extinct); *Asplenium ascensionis; Nephrodium ascensionis; Sporobolus caespitosus; Euphorbia origanoides; Xiphopteris ascensionense; Marattia purpurascens; Pteris asensionis; Dryoteris ascensionis* (or *Sporobulus durus*); and *Oldenlandia ascensionis* (now thought to be extinct). There are nine endemic marine fish, up to 26 endemic terrestrial invertebrates (Pickup, 1999) and an endemic shrimp, *Procaris ascensionis*. Ascension is 1130 km from the nearest land, the island of St Helena.

Santa Catalina Island, by contrast, is just off the coast of mainland California but has eight endemic species (one of them, Yerba Santa (*Eriodictyon traskiae*) in its protective cage, is shown in Figure 2.5). It is not the absolute distance from land that is important, but the level of isolation. Thus there is a sheep breed, the Soay, on the islands of Soay and Hirta in the Scottish St Kilda group. This separate species resembles the primitive Neolithic sheep first brought to Britain about 5000 BC and, later, onto St Kilda where they survive thanks to the isolation of these islands (Boyd, 1979). There is also a species of seaweed-eating sheep on North Ronaldsay in Orkney, about half the size of conventional sheep. Also in Orkney, on Swona Island, the cattle have been feral for four or five generations since the human population left around the late 1960s. Thus, they have not been exposed to other bovine influences, genes or to human selection. From 1999, these cattle have been regarded as a separate species, being so listed in the *World Dictionary of Livestock Breeds*.

Through their size and/or isolation, islands may lack a full set of trophic levels. The top carnivore is usually missed, which, on occasion, has seen island species react to this lack of threat. Thus, on a number of islands, birds became flightless, there being no need for escape from a predator. Another reason would be that to take flight on a small island might see a land bird blown from the island with limited possibilities of being able to return. The extinct dodo of Mauritius (*Raphus cucullatus*) is the most famous example of insular flightless-ness; others are the kiwi of New Zealand (*Apteryx australis*) and that archipelago's flightless rail, the South Island takahe (*Porphyrio mantelli*). Another is the flightless cormorant of the Galapagos that Darwin saw. Lack of competition may also see species take on unusual characteristics to exploit ecological niches not taken in the usual way.

Another characteristic of island fauna is that without (or with a restricted range of) predators, the competitive advantages of speed and agility may not be required and the species may react to a less challenged life by increasing in size. Island gigantism is fairly common as with the giant tortoises of Aldabra and the giant prehensile-tailed skink (*Corucia*) of the Solomon Islands. The Solomons also have giant rats.

The most notable case of an island having an unusual biogeography is that of Madagascar where 50 per cent of bird species, 85 per cent of plants and 95 per cent of reptiles are endemic. All the land mammals (save recent introductions), most famously Madagascar's prosimian primates, the many species of lemur, are

Figure 2.5 Yerba Santa (*Eriodictyon traskiae*), an endemic plant, Santa Catalina Island, California, USA, 1998

not found elsewhere. Sadly, Madagascar is a poor place with a high birth rate and this is reflected in pressures on the environment, which see its unique biodiversity under threat: 'Madagascar: the battle we must win', screamed a poster issued by the World Wildlife Fund (see also Nash, 1994).

Mitchell waxes lyrical about island biogeography regarding Hawaii:

> Had Darwin called at Hawaii after his historic visit to Galapagos aboard the *HMS Beagle* in 1835, he would hardly have been able to contain his excitement, for he would have seen examples of evolution which would have made those on the Galapagos pale by comparison. Tiny flies have grown into giants, daisies have turned into shrubs, crickets have gone blind, garden lobelias have evolved into trees, caterpillars have become carnivorous, geese have ceased to fly, and some small birds have even become vampires. There are numerous species of honey creepers, many brilliantly coloured but not all of them eat honey. Some prefer grubs prised from bark and have beaks for the job like woodpeckers while others have evolved to eat seeds and have tough beaks like finches.
>
> (Mitchell, 1989, p. 159)

Risks to island species

Island species, like other aspects of insular environments, are at risk. Dangers can include the direct impact of man. The dodo could not fly from and, initially anyway, did not know that it should at least try to run from the Dutch sailors when its island of Mauritius was discovered in 1598. Even if the dodo learned to run, they could not run fast enough and the species was wiped out by about 1681. A lonely stuffed dodo greets visitors to the museum in Port Louis.

It is not that continental species are necessarily safe – the extinction in 1914 of the passenger pigeon (*Ectopistes migratorius*), a North American species thought to have numbered in pre-Columbian times up to 5 billion, teaches us that. It is that island species, developing in isolation, may have even less ability to avoid new threats, be these from club-swinging sailors or novel germs and microbes.

Both island flora and fauna are at risk from introduced predators. On Ascension Island there were no mammalian predators before the island's discovery by humans. At that time the island was colonised by many tens of thousands of birds all over its lower slopes. Then, in 1815, the British took the island to garrison it against French attempts to mount rescue efforts for Napoleon from his St Helena prison. They released cats to control rats, which had come onto the island from ships. However, the cats also ate the nestling seabirds and, with abundant food supplies, rapidly grew in numbers. By 1824, residents complained of being over-run by cats, so dogs were introduced to kill the cats, although they did not become a nuisance. Within 50 years of the introduction of cats, the gannets, boobies, terns, frigate birds, etc., which had colonised the lower slopes of Ascension were gone. These species have been driven to cliffs, inaccessible ledges and an offshore island, Boatswain Bird Island, which has remained cat-free. Ascension had its own flightless bird, too, the endemic Ascension Rail (*Atlantisia elpenor*). It is possible that this bird might have survived until the advent of the cats in 1815 (Packer, 1983). Ascension's few tourists are taken on boat trips around Boatswain Bird Island and the sound, smell and sight of the dense avian life there gives them a taste of what the whole island must have been like a few generations ago. Only the island's famous sooty terns (*Sterna fuscata*), the wideawakes, remain on the lower ground, protected by an unusual life cycle. Wideawakes come to Ascension every ten months in huge numbers to breed in noisy congregation in areas of the island known as wideawake fairs. During each breeding cycle, perhaps 20,000 chicks are lost to cats and to Ascension's endemic frigate birds. However, then all the wideawakes depart and many cats starve, so there has become established a predator–prey relationship which enables wideawakes and cats both to survive on the island's lowlands. There are schemes in the offing to try and rid Ascension of its feral cats, which now number up to 900.

Ascension Island's plants are also under threat. Some have succumbed to grazing – especially from the goats which roamed the island from the early years of contact, when they were left to form a food supply for passing ships, until 1944 when they were finally eradicated. Some plants have lost their ecological niche to competition from introduced species. Around 300 alien species have been introduced to Ascension (Cronk, 1980; see also Duffey, 1964). In 1858 the Botanical Gardens in Cape Town sent 228 species; 700 packets of seeds were sent from Kew Gardens in London in 1874. The most recent problematic introduction is the Mexican thorn (*Prosopis juliflora*) which is spreading very rapidly, as explained above. It is giving the lower parts of the island a more verdant appearance, but a recent report notes that its presence is damaging to the island's conservation interest in that, potentially, it could cover 90 per cent of the land surface. This report is Ascension Island's management plan (Pickup, 1999), drawn up to help advise on its environment (see also Ashmole and Ashmole, 2000). This followed a good deal of criticism of the British approach to conservation on their island dependencies (for example, Oldfield, 1987; Pearce, 1994).

The Mexican thorn problem on Ascension is reminiscent of the serious problem on many Pacific islands caused by infestation of miconia (*Miconia calvescens*). On the South African Marion Island in southern Indian Ocean, an introduced plant, *Agrostis stolonifera* was first recorded at a base station in 1965, probably brought in as seed in fodder or imported soil. 'By 1983 it had extended rapidly across approximately one-third of the island, replacing the natural vegetation of many stream banks, drainage lines and seepage slopes' (Smith, 1987, p. 216). Another South African island, (another) Prince Edward Island, has a better conservation record in that 'only one alien species, *Poa annua*, occurs at Prince Edward Island. Strict precautions are taken whenever the island is visited, to ensure that seed is not carried on clothing, boots, helicopter wheels or camping equipment' (Smith, 1987, p. 217).

Introduced fauna, as with the case of Ascension's cats, supposedly brought in to catch the rats, have parallels elsewhere, too, such as Marion Island, where cats were introduced in 1949. By 1975 there were over 2000 and they were killing several hundred thousand burrowing birds annually. Numbers were reduced by the release of a cat virus and by shooting (Smith, 1987). On Lord Howe Island off Australia, rats were accidentally introduced in 1918. They were such a problem that, in 1930, owls were brought in to eradicate them but instead ate local birds. The owls are now a problem; the rats remain a problem and are being fought with poison. In the 1980s a programme to eradicate goats on Aldabra in the Indian Ocean began, in an effort to preserve the habitat of birds and the giant tortoises. Goats, along with dogs and pigs, are amongst the introduced species causing damage on the Galapagos Islands. Introduced goats have also caused damage on Alexander Selkirk's island of Másatierra off Chile (Wester, 1991). On New Zealand, endemic birds such as the flightless kakapo (*Strigops habroptilus*), the takahe and even the kiwi are threatened by habitat loss and predation from introduced species such as possums, deer, rats, cats, stoats and weasels. The kakapo is being bred on two predator-free offshore islands, but release into the wild on the main islands will first require better control of introduced species there. One kiwi species, the little spotted kiwi, like the kakapo survives only on an island reservation, stocked with birds taken from the mainland.

Islands are sometimes important in ecological terms because their isolation has enabled species to survive there when they have been pushed from other more accessible spots. Thus the symbol of its island's National Parks and Wildlife Service, the Tasmanian Devil (*Sarcophilus harrisii*), is now found only on the island, but once existed on the Australian mainland. Islands might also be a strategic location for migrating birds or seabirds. St Kilda, most remote of the Outer Hebrides, has a range of sub-species, of mouse, wren and, as noted above, sheep, but is also very significant for seabirds, including the world's largest colonies of gannets and the region's largest puffin colony. St Kilda is the only UNESCO World Heritage Site of the 17 in the UK to be designated for its ecology. Protection under designation or by legislation is obviously important and is to be welcomed; as with the announcement in 1999 by Australia that the government is to build the world's biggest marine park at Macquarrie Island to protect its wildlife, especially the penguins, from disturbance and from pollution. However, one cannot legislate against accidents and about 20,000 seabirds were killed in just a small oil spill in 1998 in the coastal waters of Germany's Schleswig Holstein National Park, which affected the low-lying islands of the Waddenzee. The potential impact of any oil spill from new drillings in St Kilda's region of the Atlantic would also be very serious, warned the International Union for the Conservation of Nature in 1999 (cited in the *Independent*, 6 July 1999).

Risks to islands

Finally, it should be noted that island ecosystems are also vulnerable to natural disasters. Many tropical islands are subject to hurricanes, typhoons or cyclones;

volcanic island ecosystems are threatened by eruptions. Worse than any of these is the threat to the very existence of low-lying islands. Such places are subject now to risks from sea level rise brought about by global warming, caused it is thought by the greenhouse effect. This was a major concern at the UN Conference on Environment and Development at Rio de Janeiro in 1992 and, since then, has been a focus of the UNESCO Environment and Development in Coastal Regions and Small Islands (CSI) programme. This has the aim of establishing wise practice for coastal communities including small islands. UNESCO also has an interest with the Intergovernmental Oceanographic Commission and the Permanent Service for Mean Sea Level in the *Global Sea Level Observing System* (UNESCO 1994).

Sea level rise will impact adversely all over the world on islands and mainland alike, but it is potentially at its most serious for atoll islands (Connell, 1993). The rise in sea level does not actually have to put such islands underwater to render them uninhabitable. Before that the increased likelihood of inundation in storms with damage to sea defences, infrastructure and agriculture and salinisation of the fresh water lenses would probably have driven the people away. Atolls have no higher ground to which people can relocate. In 1999 the South Pacific Regional Environment Programme reported temporary inconveniences caused by sea level rise, such as the more frequent flooding of the airport at Majuro in the Marshall Islands. There were more major problems such as salinisation of soils in Tuvalu and beach erosion in the Maldives; but worse still was the actual disappearance already of two uninhabited islands in Kiribati, Tebua Tarawa and Abanuea (cited in the *Independent on Sunday*, 13 June 1999). As early as 1990, John Connell identified the first refugees from the greenhouse effect, with people having to move from the atolls that are the Carteret Islands of Papua New Guinea to Bougainville.

> In the Carteret Islands a version of the Greenhouse effect has already occurred. Rising seas have largely destroyed the economy and forced out-migration. Islanders who were once merely economic refugees, freely moving away from their impoverished islands, have become environmental refugees. The situation of the Carteret Islanders ... is likely to be widely shared in the years to come.
>
> (1990, p. 154)

In 1998 the government of the Maldives began building an artificial island, Hulhumale, off Male, which would be able to serve as a refuge for up to half the country's population, whilst, in the meantime, also serving as an extra runway for the airport.

This chapter has shown that small islands are fragile natural systems. Their size and scale make them also problematic in physical terms for human occupation and not just because of sea level rise. Rainfall can be variable and there are very commonly water supply problems as will be exemplified in Chapter 3. The succeeding chapters will demonstrate something of the fragility of the human systems that develop on small islands.

References

Key readings

Any good general geology textbook will demonstrate the formation of islands associated with plate tectonics, those listed here by Tarbuck and Lutgens, also Wicander and Monroe certainly do. Ascension Island is much used in the text, an excellent source for its biogeography is the recent report by Pickup. Broader considerations of island ecosystems can be seen in Mitchell and Quamman.

Ashmole, P. and Ashmole, M. (2000) *St Helena and Ascension Island: a natural history*, Anthony Nelson: Oswestry.

Boyd, J.M. (1979) 'Natural history', in A. Small (ed.) *A St Kilda handbook*, The National Trust for Scotland: University of Dundee, Department of Geography Occasional Paper 5, pp. 20–36.

Cann, J.R., Elderfield, H. and Laughton, A.S. (eds) (1999) *Mid-ocean ridges: dynamics of processes associated with the creation of new oceanic crust*, Cambridge University Press: Cambridge.

Chester, Q. and McGregor, A. (1997) *Australia's wild islands*, Hodder Headline: Sydney.

Connell, J. (1990) 'The Carteret Islands: precedents of the greenhouse effect', *Geography*, 75, pp. 152–4.

Connell, J. (1993) 'Climatic change: a new security challenge for the atoll states of the South Pacific', *The Journal of Commonwealth and Comparative Politics*, 31, 2, pp. 173–92.

Cronk, Q.C.B. (1980) 'Extinction and survival in the endemic vascular flora of Ascension Island', *Biological Conservation*, 17, 3, pp. 207–19.

Duffy, E. (1964) 'The terrestrial ecology of Ascension Island', *Journal of Applied Ecology*, 1, pp. 219–51.

Mitchell, A. (1990) *A fragile paradise: nature and man in the Pacific*, Fontana: London (first published in 1989).

Mortimer, J.A. and Carr, A. (1984) 'Reproductive behaviour of the Green Turtle (*Chelonia mydas*) at Ascension Island', *National Geographical Society Research Report*, 17, pp. 257–70.

Nash, J.M. (1994) 'The making of an eco-disaster', *Time*, 21 November 1994, pp. 82–5.

Oldfield, S. (1987) *Fragments of paradise: a guide for conservation action in the UK Dependent Territories*, Pisces: Oxford.

Packer, J.E. (1983) *A concise guide to Ascension Island, South Atlantic*, privately published.

Pearce, F. (1994) 'Britain's abandoned empire', *New Scientist*, 13 April 1994, pp. 26–31.

Pickup, A.R. (1999) *Ascension Island management plan*, Royal Society for the Protection of Birds: Sandy.

Quammen, D. (1996) *The song of the dodo: island biogeography in an age of extinctions*, Scribner: New York.

Smith, V.R. (1987) 'The environment and biota of Marion Island', *Suid-Afrikaanse Tydskrif vir Wetenskap*, 83, pp. 211–20.

Tarbuck, E.J. and Lutgens, F.K. (1987) *The earth: an introduction to physical geology*, Merrill: Columbus.

UNESCO (1994) *Sea level monitoring in the small island developing states*, UNESCO: Paris.

West Wales Naturalists' Trust (n.d.) *Skomer Island*, West Wales Naturalists' Trust: Haverfordwest.

Wester, L. (1991) 'Invasions and extinctions on Másatierra (Juan Fernández Islands): a review of early historical evidence', *Journal of Historical Geography*, 17, 1, pp. 18–34.

Wicander, R. and Monroe, J.S. (1995) *Essentials of geology*, West: Minneapolis/St Paul.

3 Insularity
Processes and effects

This chapter considers some of the basic phenomena of islandness: those constraints that are imposed upon small islands by virtue of their insularity. Such matters are quintessentially geographical, insofar as small islands are affected in these matters by their location as a body of land surrounded with water which is inescapably isolated from and peripheral to the continental areas, which tend to be more significant in both economic and political terms. In addition, the chapter will introduce the rather rarer opportunities provided by insularity. As with the remainder of the book, the analysis focuses on small islands and on this topic of scale, which is the area on which problems faced by islands rest. Thus Hintjens and Newitt in their important collection of essays on tropical islands (1992) refer to scale twice in their title: *The political economy of small tropical islands: the importance of being small*. Small islands, bounded spaces, are limited in size, in land area, in resources, in economic and population potential, in political power, by their scale. There are few, if any benefits from being of small scale, perhaps exclusivity and/or privacy in respect to certain specialised insular economies, such as high-class tourism or religious uses. Usually being small scale is simply and obviously a problem (see Ólafsson, 1998).

However, there is no attempt to impose a ceiling on the size of population or area of islands to be considered here. King (1993) noted that there have been suggestions that small islands are those below 10,000 km^2 and 500,000 people (Beller, 1986) or, alternatively, below 13,000 km^2 and one million people (Dolman, 1985), but such distinctions are artificial. King found them 'not particularly helpful' for the Mediterranean (1993, p. 17); nor would they be anything but frustrating if imposed on small island studies elsewhere. Thus in the Indian Ocean, Beller's threshold regarding population, if not area, would exclude Réunion and Mauritius from consideration, although both are framed by their insularity (though see Eriksen, 1993, for a discussion of this idea applied to Mauritius). Even more difficult would be Dolman's definition. This would count Réunion as a small island, given its population of c. 624,000, but not its neighbour and direct comparator, Mauritius. The latter is slightly smaller at 1865 km^2 (for the main island, 2040 km^2 for the state) to 2510 km^2 for Réunion, but has a larger population of 1.1 m, above Dolman's threshold. The approach here is just to acknowledge that the effects of insularity tend to be more pronounced the smaller the island is, but islands of any size will be brought into the discussion where appropriate. Chapter 8 will be a detailed consideration of the problems of scale.

In addition to scale, small islands face specific problems with isolation, which impacts on accessibility to services and to markets as well as bringing danger and inconvenience in the necessity to travel over, on or under the water surrounding the islands. Isolation, together with scale, often distance islands from political power. By contrast, there are occasions when isolation can be a resource. However, sometimes it has not been islanders who have profited in such cases; a sad story will be told of the way in which islands and islanders have been sacrificed for the benefit of mainland states/colonial powers with regard to atomic weapons testing. This brings in two current concerns in human geography, contestation and the experience of the other. Islands required for significant activities, such as weapons testing, will be taken however much the locals, the other, contest the activity. Island histories are replete with examples of military, political and economic domination from outside, as will be demonstrated below. The chapter will conclude with an introduction to island economic matters with a consideration of resources.

Isolation and peripherality

The problems of isolation

To revert to the frustratingly simple dictionary definition of Chapter 1, an island is a body of land surrounded by water. As such, to get people or goods onto or off an island requires that body of water to be crossed in some manner – by boat, plane, via a bridge, causeway or tunnel, or by swimming or wading. This is the most obvious and basic constraint of insularity. This water crossing has to be an inconvenience at the least. Usually it is also an expense, one has to pay for the journey either in terms of obtaining a place on a ferry or plane, or through taxes, or by toll in paying off the capital investment that provided a fixed link. Pupils from the Scottish Western Isles (Outer Hebrides) regularly journey to see Shakespeare performed in the theatre in Inverness. They travel up to 30–40 km from their homes to catch the 5.30 a.m. ferry from Stornoway. Then they endure the three-and-a-half hours crossing to Ullapool; travel by bus 80 km to Inverness and hang around until the evening performance. They spend the night in a Youth Hostel before arising at 6 a.m. for the return journey (J.S. Grant, letter in the *Independent*, 3 March 1998). Shakespeare with seasickness pills, such are the impacts of insular isolation.

Small island producers are especially affected by isolation. Take the case of agriculture. In some rare cases an island, because of its environment or other circumstances, can become an important supplier of a particular crop. One example was nutmeg, originally grown only in the six tiny Banda islands, 1000 km north of Australia. Run Island was the main supplier, and the location of this island was a closely guarded secret in Europe, to where most of the nutmeg was traded via Asia. Then, in the sixteenth century, Portuguese sailors landed on the Bandas, setting off contestation between European powers for control of the islands and the lucrative nutmeg trade. A British adventurer held off the Dutch from Run for four years from 1616 before losing the island and his life. Later, a treaty legitimised Dutch ownership, granting Britain, in return, legal possession of another island, one seized from the Dutch in 1664. This was Manhattan Island. This was a measure of the importance of Run and the other Banda Islands in the sixteenth and seventeenth centuries (Milton, 1999). In the nineteenth century the British successfully transplanted nutmeg to Ceylon (Sri Lanka) and Singapore, larger islands where economies of scale saw production far exceed anything the Banda Islands could match and the islands returned to obscurity, their economies destroyed.

Usually, island produce is not distinctive and producers such as farmers have to compete in at least a regional if not a world market with mainlanders generating similar goods. Say that product is sheep (Figure 3.1); they will be matched directly against sheep from the mainland in the market place. Unless there is something very special about them (as with the North Ronaldsay sheep which are marketed separately), there will be no extra premium; they are just sheep, and the island farmer will receive the same sort of price as a competitor from the mainland. Yet from that standard price the island farmer has to fund a considerable number of additional costs. These might include hiring a boat, or, paying for space on a ferry, or, if there is such an option, of paying for a lorry to come to the island to get the sheep to market. The journey with the sheep might of itself be troublesome; in 1986, 5000 sheep from Coll and Tiree jumped ship in

Figure 3.1 Sheep being exported from Clare Island, County Mayo, Ireland, 1984

the Scottish mainland port of Oban, causing some chaos and confusion before all were recaptured. The island farmer will have had to pay also for freight charges to bring over any foodstuffs bought in or fertilisers, pesticides, seed, whatever may be required for the grassland, as well as equipment for the farm. If he or she needed to send for a vet, unless the island is large enough to sustain its own practice, the vet will have had to make the sea crossing and these costs will doubtless find their way onto the farmer's bill.

Sometimes, in addition to cost there is danger in travelling to or from an island. It is a not infrequent occurrence to hear of ferry or other boating disasters, even in the modern era. In December 1999, the *MV Asia South Korea* sank between Cebu and Iloilo in the Philippines, fortunately the considerable majority of the 650 people on board were rescued. In September 2000, 79 people died off the Greek island of Paros, when the ferry, *Express Samina*, struck rocks and sank. In December 1998, the tiny island of Iona off Mull in the Scottish Inner Hebrides lost nearly six per cent of its population when four young men between 19 and 24 years old drowned when their dinghy overturned on their return to Iona after attending a dance in Mull. Only one man survived. The Iona community, already hit by the planned closure of its school, was devastated by the loss of the only men of their generation who had decided to make their life on Iona rather than migrate. Nor is it just boat journeys that can end in disaster. In 1996, 141 people were killed on Spitzbergen (Svalbard) off Norway when a chartered jet carrying Russian miners and their families crashed into a mountainside.

Most journeys on and off islands, of course, are made without incident. Further, the very isolation that necessitates the special journey is important in Baum's 'fact of difference' (1997) which is part of the appeal of islands. As such, this can be important in the tourist trade, but it cannot be gainsaid that having to cross the water to an island is a direct penalty imposed by insularity. (See Hamilton-Jones (1992) for a study of this issue in relation to inter-island shipping in Fiji and the Cook Islands.)

Other penalties of isolation may be less obvious but can still be important. Islands' isolation makes them peripheral. In the case of continental countries with islands, these have perforce to be on the edge of the country, usually far in absolute and social terms from the major central places, such as the national capitals. Most continental countries with coastlines have islands (Belgium is one of the few exceptions) and in all but two cases their capital cities are on the mainlands. Some capitals originated on coastal islands, usually for defensive purposes, but such settlements have long been incorporated into a principally mainland city as with Lagos, the former capital of Nigeria and Stockholm in Sweden. The two continental countries with capitals wholly on offshore islands are Denmark and Equatorial Guinea. In the Danish case Copenhagen is on an island in the extreme east of the country; one which only in the 1990s was connected by fixed links to the mainland of Europe. The logic of this peripheral insular city becoming the capital can only be appreciated by consideration of Denmark's history. This is helped by an understanding of Copenhagen's place name derivation: it is a corruption of *Køpenhavn* or 'Merchant's Harbour', this explaining its coastal location. It actually became a major trading city when Denmark and Sweden were united and if both countries are considered together, Copenhagen has a strategic central location. By the time Denmark and Sweden separated, Copenhagen was already a dominant city, having eclipsed formerly important towns on the Danish mainland, the Jutland peninsula, such as Ribe (Court, 1987).

The case of Equatorial Guinea is not dissimilar. This country is made up of the former Spanish colony of Rio Muni (now called Mbini) on the mainland and five offshore islands, of which the most important are Bioko (formerly Fernando Po) and, far to the south, beyond the insular state of São Tomé and Príncipe, Annobon (once Pagalu). Of these areas of land, the most significant and developed was Fernando Po, as this island was used as the Europeans' base in this area (Lynn, 1984). Being offshore, Fernando Po was both more easily defended and healthier than the mainland – West Africa being feared, with justification, as the 'white man's graveyard'. The Spanish did not really penetrate much into the Mbini hinterland until the 1920s; Fernando Po, by contrast had been exploited and developed by Europeans since the fifteenth century. Upon independence in 1968, when these areas of Spanish land were coalesced into one, troubled country, the city of Malabo (formerly Santa Isabel), with its fine natural harbour on what had been Fernando Po, had the best infrastructure and was the logical choice for the capital. Elsewhere, countries made up of continental and island areas, even those such as Greece where the island realm is very significant, have their capital cities on the mainlands where there is more and easier connectivity to the rest of the country and the outside world.

Capitals are usually centres of innovation and certainly the locus of political power. Areas remote from the capitals inevitably lose out. Being away from the capital and from other big cities can make places in truth or perception relatively backwards. This is the reputation, doubtless incorrect in reality, of the mid-west of the USA at the heart of a great continent, but remote from the vibrant coastal cities. How much more is this physical and social disconnection a trial for small islands, isolated in every way from the heart of their nation? Any rural people may be looked down upon by oh-so cultured and up-to-the-minute folk from the big city – this is known in Ireland as 'jackeens' versus 'culchies', with the (self-declared) superior Dubliners being the jackeens and the rest being culchies. Islanders, being from even beyond the rural periphery are inevitably the equivalent of culchies in any mainland/island state. This is evident, for example, in the French State where the islanders can be thousands of kilometres from the French mainland. 'There is widespread resentment of many islanders ... at the day to day discrimination and prejudice from metropolitan French people, both in the islands and on the mainland' (Hintjens, 1992, pp. 64–5). A recent newspaper article was entitled: 'The people of Orkney don't speak Gaelic (and never have), do have electricity and their eyebrows don't join in the middle. But could a mainlander tell you that?' (*Independent on Sunday Magazine*, 20 September, 1998). Such dismissiveness is usually just an irritation to island culchies, if an irritation of long standing – recall Cummian's seventh-century dismissal of the Columbans on Iona as living on a pimple on the face of the earth, mentioned in Chapter 1. There could be a real difficulty, though, in that such reputations, however unjustified, may make the peripheral islands less attractive to inward investment. This is a problem to add to the disincentives already present in the higher operating costs to be found in an island location and the distances, social and/or physical, from major markets. For generations in long-settled developed regions such as Europe, islanders have reacted to their peripherality by migrating, either to the New World or to the cities (see the essays in King and Connell, 1999, also Chapter 5 of this book). There are more people with knowledge of Scots Gaelic in Glasgow now than in the language's heartland in the islands off Scotland's West Coast; there are more people from the Azores in Toronto than in their home islands.

In island nations the urban/rural; core/periphery distinctions might be expressed in relationships between the dominant island and the off-islands. In Singapore, by far the most traditional settlement, really the last surviving rural village is found on an island, Palau Ubin, off the northeast coast of Singapore Island itself. On the main island such village forms have all been replaced by new town-type developments. Indeed some 'traditional' villages have now been recreated as museum pieces for tourists and heritage purposes, as on the playground island of Sentosa off Singapore Island's south coast. On isolated Palau Ubin there is still a real Malay village (Figure 3.2). Singapore's authorities in recent years have become conscious of having sacrificed too much of their traditional built forms to modernisation and so this last village will remain, unless and until Singapore's population grows to four million from its present 3.2 million. Upon this the plan is for Palau Ubin to be connected by causeway to the main island and be subject to complete rebuilding.

Figure 3.2 Malay village, Palau Ubin, Singapore, 1995

The state of Mauritius also has an insular periphery, in the form of the island of Rodrigues, 554 km eastwards, far smaller in terms of both area (104 versus 1865 km²) and population (c. 35,000 versus 1.1 m) and much less developed. Many islanders from Rodrigues have moved from the periphery to the centre and now live in Mauritius, some in distinctive communities on the poorer outskirts of Port Louis. Mauritius State is two distant islands joined together politically; other insular states are true archipelagoes, such as Cape Verde in the Atlantic, the Seychelles in the Indian Ocean and many others. In such cases the capital island can become overwhelmingly dominant. In the Marshall Islands, Kiribati and Tuvalu, the capital islands (actually atolls) of Majuro, Tarawa and Funafuti respectively have the international airports, the seats of government, the major urban centres and act as the chief central places for their nations. Their off-islands may not be dissimilar in size and basic resources to the capital islands but become relatively underdeveloped, with fewer opportunities available to their populations. The people may migrate to the capital islands as a result. The seriousness of the problems such emigration causes to the outlying islands in terms of a loss of labour and an ageing population might perhaps be matched by the difficulties immigration causes to the capital island in terms of overcrowding and pressures on the environment and infrastructure. In the report of the 1978 Kiribati Census, there appeared a perceptive quotation from A.V. Hughes, an ex-patriate government economist in the days when Kiribati was part of the British Gilbert and Ellice Islands colony:

> Urbanisation is a problem only because the people urbanising make impossible demands . . . on the physical resources of the town. The allocation of more physical resources usually only makes things worse by accelerating the inflow of people who make use of them. . . . Taking the national view of resource allocation, urbanisation is seen as de-ruralisation. People leave social and physical resources in the rural areas (which then decay) to go to non-existent social and overstrained physical resources in the town.
>
> (Bailey, 1983, p. 84)

Peripherality is, of course, a relative concept but the problems it brings are absolute. Thus the Falkland Islands are very peripheral at a world scale. However, within the archipelago, which as a whole is perhaps the least densely populated habitable place in the world, there are variations in this peripherality. The one town of Stanley on East Falkland is overwhelmingly dominant with about 1200 of the 2000 population and there has been considerable migration from the rest of the islands (known as Camp) into Stanley in the 1980s and 1990s. Since the Falklands Conflict of 1982, the already ongoing process of land reform that was leading to the break-up of big company estates into family farms accelerated. The shake out of non-familial labour that accompanied this process has reinforced migration from Camp to Stanley, to the extent now that there are concerns for the continued viability of some of the peripheral rural settlements on West Falkland and some of the smaller islands (Royle, 1994; Dodds, 1998).

In sum, isolation, whether relative or absolute, is a problem that is faced by island populations. There are circumstances, however, when isolation has proved to be an advantage, as the next section will show.

Isolation as a resource

Getting away from it all

We all need to 'get away from it all' occasionally. For most of us this means trying to find personal time to relax and unwind, anything from a hot bath to the annual holiday. However, some people need to be away from it all permanently. In New World countries such as the USA, groups wishing to live a secluded religious life, such as the Mormons, could find isolation by going to the frontier, to Utah in their case when they were faced with persecution in their previous home in Nauvoo, Illinois in the 1840s. Another, smaller, group, the Amana people, who lived a strictly communal life, fled from their original home in Germany to the New World, to Buffalo, New York state in 1842. They were able to flee again to the eastern part of Iowa, then on the frontier, from 1854 when their isolation was again compromised. Once the frontier filled up, when faced a third time with pressures from the outside world there was no longer anywhere left to go and the communal society broke up in 1932. By contrast in Europe, which had long been fully settled, groups seeking isolation had often to find it on islands. Holy Island (Lindisfarne) in northeast England, Holy Island off Aran in Scotland, Iona, Skellig Michael off western Ireland, Mont St Michel in Brittany, St Michael's Mount in Cornwall, all these places had/have religious connections – their isolation became a resource which was utilised by special groups. Off the coast of south Wales, a monastery owns and occupies Caldy Island. There are regular ferry links with the mainland, the monks have to have an income and the tourists that come in large numbers in the summer are a ready market for the perfumes the community makes from the island's flowers. But every day the last ferry leaves; access and numbers are controlled and the monks are able to spend much of their time in quiet contemplation in a way that would not be so easily achieved in a mainland setting. The island is a refuge for the soul again.

The utility of the island refuge is not solely a Christian tradition; thus, the isle of Pemba, now in Tanzania, was the site for black Africa's first mosque from the eighth century when a dissident sect fled from what is now Oman after a dispute over doctrine. There are even multifaith island refuges, thus Holy Island off Aran in Scotland was an ancient centre for Christianity, but in 1994 was bought up by a Buddhist group.

Islands as stepping stones

The concept of the island as stepping-stone will be dealt with more fully in Chapter 4, as it is more of a historical than a contemporary phenomenon. The idea is that an island, being small and bounded, can serve as a secure base from which interaction with a larger continental area can be occasioned. For centuries, Arabs, and, later, Europeans used Zanzibar as a base from which to launch incursions into East Africa, be they for trade or slaves; on the opposite coast, Fernando Po (Bioko) served a similar function (Lynn, 1984; 1990). In the USA, in the huge stream of migration from Europe at the end of the nineteenth and start of the twentieth century, most migrants were processed through Ellis Island in New York harbour (Figure 3.3). This location ensured that they could not gain access to the continental mainland until they had been properly approved and checked for disease. Stepping stones can be used when leaving a continental location, too. For example, the nationalist Chinese took Taiwan when driven from mainland China by communist forces in the 1940s. Also connected with migration to the Americas, a much grimmer aspect of it, is Goree Island off Senegal where slaves were kept secure before being forced through the 'Gate of No Return' on to the ship which was to take them into bondage an ocean away. United States President Bill Clinton, in his tour of Africa in early 1998, was taken to this island and seemed suitably contrite.

Islands as prisons

Another island to which President Clinton went during his African tour was Robben Island, 11 km off Cape Town. He was accompanied by a fellow president on his trip there, Nelson Mandela of South Africa. Mandela had been an unwilling resident of the island for many years, Robben Island being the prison where he and many other black activists were held during the apartheid era. It is now a National Heritage and Conservation Site and has been subject to application to become a UNESCO World Heritage Site. Robben Island's earlier uses were also predicated upon its isolation: a whaling station, a leper colony and a lunatic asylum.

Devil's Island off French Guyana is another notable example of a prison island. Others include Spike Island, Ireland and Rottnest Island off Perth in Western Australia. St Helena is a famous prison island, as will be discussed in Chapter 10. It is problematic leaving islands at the best of times, to have to escape from not just the prison building but then the island and its stretch of water is difficult indeed. Within a few kilometres of central San Francisco is the island of Alcatraz, formerly used as a US federal penitentiary, most notably for gangsters such as Al Capone. Nobody for certain ever escaped from Alcatraz, which is protected by its cold waters and strong currents. Those few who made it off the island almost certainly drowned.

Some islands were penal colonies. The Falkland Islands were considered for this purpose in 1840; Tristan da Cunha had been similarly proposed in 1785; Australia, the island continent, was used instead. The Australian penal colony system was actually quite complex and made use not just of the Australian mainland, but of several offshore islands for prisoners deemed necessary of especial incarceration. Sarah Island, Melville Island, Stradbroke Island, Norfolk Island

Figure 3.3 Ellis Island, New York, USA, 1992

and Cockatoo Island were used for these purposes (Pearn and Carter, 1995). Cockatoo Island, under 10 km from Sydney Cove, but protected by deep water, was one of a number of islands in Sydney Harbour which were used as penal settlements. This was from 1839 to 1871 for male 'convicts of the most desperate and abandoned characters' (O'Carrigan, 1995, p. 65) and the buildings then had a second lease of life as a female penitentiary until 1908. Tasmania was also used, particularly the Tasman Peninsula, almost an off island, connected only by a narrow neck of land that could be closely guarded using dogs. On the peninsula was built the notorious penal settlement of Port Arthur, developed from 1830 and used until 1877 (Brand, n.d.). A flavour of the horrors of life there can be gained from reading what must be one of the bleakest novels ever written, *His natural life* by Marcus Clarke (1875). Port Arthur and other penal institutions in Tasmania are now the basis of a tourist industry (Young, 1996), as are the facilities on Rottnest Island, Western Australia, which was used as a prison for aborigines from 1838. The first prison burned down in 1856 and was replaced by a remarkable octagonal building in 1864 amongst a range of prison buildings and facilities, including a reformatory. The reformatory closed in 1901 and the main 'native prison' in 1903. The island was then developed as a health resort, but also retained some institutional use as an army camp and housed some (white) prisoners. Prisoners of war were held there in both World Wars. Today the island is a tourist resort and the insensitive can even board in prison buildings once used to house aboriginal people found guilty of transgressing laws imposed upon them by the invaders of their land (Ferguson, 1986).

Strategic locations

Islands can, on occasion, be of temporary or permanent utility to outside groups because their location is strategic in some way. The islands as stepping stone idea has already dealt with some aspects of this: thus, Zanzibar gave defensible, strategic access to East Africa. Strategic locations are usually conceived of in military terms, however, and islands throughout history and around the world have fulfilled this role. Bermuda in the north Atlantic has long prospered from its location relatively close to the East Coast of the USA; sometimes the relationship has been purely commercial, as with today's tourism activities and the early twentieth-century period when the island provided vegetables for the USA market. At other times, the island's location proved strategically valuable in the usual military sense. Thus in the war of 1812 between Britain and the young USA, it was very useful for the British to own an island – strictly speaking an archipelago – with excellent harbour facilities a long way from home and relatively near to the enemy shore. In the famous raid on the White House in 1812, the British ships involved were grouped together and made ready at the British naval facilities in Bermuda.

In the days of sail, holding an island with access to a potential rival's shipping lanes was of considerable importance. The history of the West Indies is laden with examples of strategic islands being used in the interminable wars between the European powers. In the Mediterranean, the British and the French squabbled over possession of Menorca because of its fine harbour and strategic location near the Straits of Gibraltar. Menorca was British from 1713–55, from 1763–81 and from 1798–1808, after which it was no longer contested by them and it went to Spanish control. The island has some legacies of its colonial past. There is a local liking for gin. This and the only sash windows in Spain are testimony to British rule, whilst, fittingly, France's main legacy is a culinary one, a sauce created first in the island's city of Mahon, and passed down to us as mayonnaise.

In World War II, islands could become stationary aircraft carriers and dozens of islands in the Pacific and elsewhere date their airstrips to structures hurriedly built by the Americans in the war. The Japanese, too, built airstrips on their Pacific island possessions. Sometimes the island was little bigger than the airstrip, as at Taroa Island, part of Maloelap Atoll in the Marshall Islands. The Marshalls and other Pacific Islands proved their strategic value in this war as some of the bloodiest battles in the whole conflict were over possession of tiny scraps of land, whose sole utility was their ability to provide storage bases, shelter for shipping and landing strips for aircraft. The *Enola Gay* took off from a captured Pacific island, Tinian in the Northern Marianas, to drop its atom bomb on Hiroshima; Tinian was a strategic asset indeed. As was Cuba to the USSR – the world came close to Armageddon in 1963 when the Russians put missiles on the territory of their insular ally Cuba, strategically positioned near the coast of Florida.

In the modern period, long-range aircraft and telecommunications facilities have rendered the strategic value of some islands of less utility and many military facilities have been closed. Thus the Americans gave up their bases on Bermuda in 1995 and allowed themselves to be expelled from the Philippines in 1992. The Americans retain a global presence, but now manage their global affairs from far fewer places, which still include a number of strategic islands. Guam in the Pacific is one. This island has been separated politically from the rest of the Northern Marianas, so it can remain American territory without the inconvenience of the US base being placed on foreign soil, with foreign sensibilities to be offended. Thus the people and government of the Japanese island of Okinawa have expressed growing bitterness about the continued presence of American bases there, so long after the end of World War II. A non-binding plebiscite in April 1998 voted overwhelmingly for the Americans to leave. The Americans continue to use, where necessary to their purposes, strategic assets of other countries. For example, there is an American missile testing facility at Kwajelein Atoll in the Marshall Islands. In the Indian Ocean, the Americans have a huge base at Diego Garcia, theoretically British as part of British Indian Ocean Territory. However, the British are a particularly strong ally of the Americans and allow the superpower to use Diego Garcia. The British even removed people from their homes lest they got in the way of American operations. British Indian Ocean Territory is officially uninhabited now, except for service personnel. Meanwhile, the displaced islanders, *les Illois*, live in some poverty in Mauritius, a fact that should shame the British authorities. This abuse of islands and islanders for the greater purposes of colonial powers is a topic to which we shall return.

In the Atlantic, too, the Americans retain a strategic airbase and missile tracking station on British territory on Ascension Island. Allegedly, they were unhappy when the British wished to use the island and its airstrip to a greater extent during the Falklands Conflict in 1982. RAF Ascension became a very important strategic asset, a forward supply base for the British expeditionary forces which went down to the Falklands. In fact, Ascension Island is only a strategic base; it has never been allowed to develop a resident population since the British first put a garrison on the island in 1815. Only in 1999 did the British Foreign and Commonwealth Office commission a report that was to consider how to bring a normal civilian society and economy to the island. All people on the island are workers, their families or passengers in transit to/from St Helena and the Falkland Islands, Ascension being a port of call for the ship which is the sole transport link to St Helena, which has no airport. Ascension is the refuelling base for the Royal Air Force jets, which provide both the military and civilian air services from Britain to the Falklands. Ascension's variety of roles throughout history have always been predicated on its convenient location as a speck of dry land in a huge expanse of ocean; it is actually a volcanic product of the mid-Atlantic ridge. At first, in 1815, it was taken as a base by the British to deny its possible use to the French to mount a rescue bid for Napoleon, imprisoned on St Helena. Later in the nineteenth century it was used in connection with British interests in West Africa, as a base from which to mount bids to capture slave ships when the British, finally, were actively suppressing that dreadful trade, rather than participating in it. From 1899 Ascension was used as a landfall for trans-oceanic communications cables for the Eastern Telegraph Company, later Cable & Wireless. In 1942 its airstrip was built by the Americans to help them move men and *materiel* to Africa. From the 1950s it became a missile tracking station for American intercontinental ballistic missiles; from the 1960s a telecommunications base. Thus, the BBC broadcast World Service programmes from there, although now the corporation has subcontracted the actual process of transmission to another company, Merlin Communications. The National Aeronautic and Space Administration used Ascension as a tracking station during the Moon landings; Cable & Wireless now administer a base from which the European Space Agency tracks the Ariane rockets launched from French Guyana about every three weeks at present. Ariane rockets are launched eastwards across the Atlantic, for about six-and-a-half minutes of each of the brief but hugely expensive flights the rockets can be tracked only from Ascension Island.

Islands as places far away ...

Isolation can have further utility in that islands might be used for activities that are not suitable to being carried out in heavily populated regions. Thus the Americans are using their Pacific possession of Johnston Atoll, one of the remotest places in the world, to dispose of unwanted munitions from the end of the Cold War, including the American nerve gas arsenal once stored in Europe. Biological and chemical weapons can be made safe there in the knowledge that should something go awry with the process, there are few people nearby to be harmed, although of course the marine environment would be at risk. This atoll has been a secret US base since the 1950s and was used for atmospheric testing of nuclear weapons in 1962 (Table 3.1). In 1942 and 1943, during World War II, the British used Gruinard Island in the Hebrides to test the environmental impact of anthrax, including the effects on sheep tethered downwind. The island was subject to expensive clean up procedures from 1986 after it was determined that the anthrax spores were not breaking down environmentally as had been hoped.

The use of islands for such matters has not always been free from abuse or controversy. The saga of nuclear testing is an example here, when two factors associated with insularity – isolation and powerlessness – combined to permit islands and the interests of islanders to be sacrificed to meet the military needs of more powerful actors. For example, in 1946, the Americans began Operation Crossroads in the Marshall Islands, then a component of the Trust Territory of the Pacific, administered by the USA. To be fair, it is recorded that Chief Juda, the *iroij*, or leader, of Bikini Atoll, one of the two atolls used for US weapons testing, did give his consent for his atoll to be used from 1947, for the 'benefit of all mankind'. However, it is doubtful that Juda, leader of 161 Micronesian islanders, would have been able to stop the Americans doing just what they

Table 3.1 Islands used for nuclear weapons testing 1945–98

Island	State	Number of tests
Mururoa Atoll	French Polynesia	174
Enewetok Atoll	Marshall Islands	43
Christmas Island (Kiritimati)	Kiribati	30
Bikini Atoll	Marshall Islands	23
Johnston Island	USA	12
Fangataufa Atoll	French Polynesia	12
Malden Island	Kiribati	3

Source: *Bulletin of Atomic Scientists* at http://www.bullatomsci.org/research/qanda/tests.html

Notes
1. 24 different places were used for nuclear weapons testing from 1945–98. Seven were islands (29%).
2. There were 2051 nuclear tests known about between 1945 and 1998. 297 took place on islands (14.5%).

wanted. The Americans were the controlling power in whose 'trust' Bikini (and Enewetok, the other atoll used) had been placed.

Operation Crossroads was the testing of nuclear weapons in the atmosphere and in all between 1946 and 1958 when the practice was outlawed, 66 nuclear explosions were carried out from Bikini and also from Enewetok atolls. A test at Enewetok in 1952 reduced the 40 named islands of the atoll to 39 as Elugelab was totally vaporised. The most powerful explosion was Bikini's Operation Bravo in 1954, 100 times more powerful than the Hiroshima bomb, and this polluted about 50,000 square miles of ocean. Three atolls and a Japanese fishing boat were badly affected, showered with radioactive pulverised coral from the explosion, 'Bikini snow'. Many innocent Marshallese on Rongelap suffered radiation burns and other problems, unto generations unborn, but were not taken off their polluted island for 48 hours. Perhaps more than 20 atolls were affected to some extent. The 253 Rongelap people were returned to their island in 1957 and have suffered continued health problems, for example 75 per cent of those who were under ten in 1954 have had surgery for thyroid tumours. The Americans declared that Rongelap was safe, though shellfish, as filter feeders and accumulators were not to be eaten. In 1985, the residents accepted the offer from Greenpeace to be relocated to Mejato, one of the islands of the huge Kwajelein Atoll.

The residents of the two atolls used as the site of the explosions, unlike those of Rongelap and also Utrik where there were also problems from Bikini snow, were evacuated prior to the tests. The Enewetok people were relocated to Ujelang Atoll. After a $120 m clean-up programme to deal with the effects of the 43 bomb tests there, they were allowed to return to Enewetok Island, about 18 km from Runit Island at the other end of the atoll where contaminated material is stored under a concrete dome. The 161 residents of Bikini Atoll were moved prior to the 23 tests that took place there, first to Rongerik, then to Kwajelein and then to Kili, moved for the greater good of the American government under whose trust they lived. In 1969 some were allowed back to Bikini. Tragically, this was a wrong decision and people became contaminated by cesium ingested from food grown on Bikini. They were moved off once again, in 1978, and now live in Ejit, an island of Majuro Atoll. A guidebook to Micronesia notes that they are protective of their privacy and visitors are discouraged from wading out to that island (Bendure and Friary, 1995). A plan to extend the atoll's road through Ejit to islands beyond is not popular with the Bikinians. Meanwhile, the Americans are cleaning up Eneu and Bikini islands in Bikini Atoll, under a $42 m plan of 1985 which would encourage the US Congress to back the political change in the Marshall Islands' status from trust territory to its present independent status in free association with the USA. In 1996 Bikini was opened up for diving tourism to the ships sunk there during the tests. There is a proposal for the Marshall Islands, where the economy is currently very weak, to accept nuclear waste for storage on still-polluted islands at the opposite end of the atoll from Eneu. Needless to say, this is controversial in the islands.

Elsewhere in the Pacific, a not dissimilar saga though without so much damage to humans unconnected with the military can be traced through the actions of the French and British. The British conducted most of their atmospheric tests in the Australian outback but also in 1958 used a British colonial island, Christmas (Kiritimati) Island, in what is now Kiribati, on the same principle of remoteness. The French used the remote Polynesian atolls of Mururoa and Fangataufa, although from 1960 to 1966 they had used part of the Sahara desert, then under French control. The French point out that there are only 2500 people living within a 500 km radius of the Polynesian sites, as opposed to several million within a similar range of the American facilities in Nevada. After 45 French tests, atmospheric tests were outlawed, and the atolls were used for 134 underground tests before 1992 when a moratorium was imposed. The explosions actually took place deep within the volcanic rock underlying the lagoon rather than within the superficial coral deposits that make up the actual atolls themselves. In 1995 the French resumed testing in the Pacific. This saw the French subject to worldwide criticism: 'World unites to condemn French' was one South Pacific headline (*The Australian*, 2 July 1995). 'We can live for six months without being liked by the New Zealanders' was reportedly one official's response to the protests (Usher, 1996), but more troublesome might have been a drop in sales of French goods. In 1996, the work done, the bomb presumably perfected, the French government promptly signed the nuclear test ban treaty. Mururoa Atoll had been subject to 174 tests and Fangataufa, 12 (Table 3.1). Whilst one would not ideally choose such an economy, the nuclear testing actually employed many thousands in French Polynesia and the ending of the test programme caused economic hardship in the territory. There was a transitional period in which the French made substantial funds available to stimulate economic diversity in Polynesia. This story of nuclear testing is just an extreme case of the powerlessness of islands, other aspects of which will be examined in the next section.

Powerlessness

Small islands are places without power. On occasion, as when being used as a stepping stone, an island can have a dominant role over a mainland area, but this does not usually last. Thus Zanzibar is now just a somewhat unwilling off-shore part of Tanzania, ruled ultimately from Dar-es-Salaam (Robinson, 2000).

All islands have been in a position of political subservience to an outside power; most still are, even if that political subservience may now be expressed in neo-colonial rather than colonial terms. No island has always maintained its political independence. Even Great Britain and Japan, the two major insular powers that themselves held huge empires, have passed periods under foreign rule and influence. Great Britain was subject to invasion and colonisation from Romans, Danes and Anglo-Saxons. Admittedly this was many centuries ago but the islanders in Britain could then not hold out against superior outside forces from the continent any more than could the Grenadians stop the Americans invading in 1984. And let one recall that southern England was very close to being invaded in World War II, until Hitler turned his attention instead to the vain attempt to better Napoleon and mount a successful invasion of Russia. He failed; he should have invaded the island.

Japan has almost always been ruled by Japanese, and often the archipelago held itself aloof from the world, in its insular fastness. However, this island nation did become heavily involved in World War II, initially, of course, by mounting a devastating attack on an island target, Pearl Harbor in Hawaii. Later, the Japanese were beaten back from many Pacific islands they had taken by colonisation or conquest including some of the Marshall Islands and some Gilbert Islands. The defining battle was on Okinawa, the only Japanese home island to be invaded; though it is far to the south of the main Japanese archipelago. There is no doubt that the Americans' insular invasions would have continued up to Honshu if necessary, but the alternative strategy, of dropping atomic bombs on Hiroshima and Nagasaki, rendered further invasions unnecessary and the Japanese surrendered. The Americans, under General MacArthur, then ruled Japan for a number of years.

Every other island is either an offshore part of a mainland state, or, if now part of an independent insular nation, was once a colony, or at the very least, like Tonga or Bahrain, a protectorate of a colonial power. Usually these were European, but the Americans, Australians and New Zealanders have also been and, indeed, still are responsible for the governance of distant islands. We see here the operation of unequal power relationships. In politics, particularly when politics descends into conflict, might is all that matters and usually, the more powerful protagonist wins. Continental but tiny Kuwait could not resist the invasion by Iraq in 1990; Iraq though much larger, then could not resist the more powerful multi-national forces arranged against it, led by the Americans. If battles are for islands, the islands, because of their scale, again, are rarely the more powerful protagonist. It may be that their isolation may help, that the invaders from afar may struggle to get sufficient forces assembled off the island, that they may be beaten off, but if the larger power wants to take the island enough they will return with more forces and succeed. In a very prominent position in central Palma in Mallorca is a statue of a warrior with a slingshot. This represents the triumph of the local people who fought off a Roman invasion fleet through the actions of their soldiers who were armed with slingshots (these could propel a projectile the size of a fist with surprising velocity). What happened next? The Romans returned in 123 BC with more ships and had built on them barriers beneath which their soldiers could duck, protected from slingshots until the ships could force a landing. At close quarters the Roman swords were the superior weapons and the Romans conquered Mallorca. When the empire collapsed, the Vandals took the island in AD 465; the Byzantines took it in their turn in AD 533, then the Moors from AD 825. They were expelled by a Christian invasion in 1229, issuing in a period of independence when Mallorca had its own kings. However, the island was absorbed into the mainland kingdom of Aragon in 1340 and went on to become part of Spain, although it was contested unavailingly by the deposed Mallorcan King at the Battle of Lluchmayor in 1349. Palma's slinger is in the shadow of the Almudaina, a building dating back to the Arab period; next to that is the magnificent Gothic cathedral, both buildings representing cultures alien to the native slinger. Mallorca's political powerlessness continued, it was bombed during the Spanish Civil War and at its conclusion, there was some discussion about the island being given to Italy, as a reward from Mussolini's help to the victorious forces of General Franco.

Only rarely has an island fought off a determined invader. The prime example took place in 1565 with the Great Siege of Malta. Malta had been given (in an example of island powerlessness) by Holy Roman Emperor, Charles V to the Knights of St John in 1530. This multi-national Christian force had originally been founded during the Crusades but had been forced from the Holy Land, and ended up, in a stepping stone idea, but stepping back this time, on the Aegean islands of Kos and Rhodes. They had been defeated there in 1522 by the Turks under Suleiman the Magnificent and had been without a permanent home until gaining Malta. In 1565, the Turks, still ruled by Suleiman, determined to have a defining battle with the knights and to take the island of Malta, which, with its strategic central Mediterranean location, would have made a useful base for further conquests. Suleiman sent an overwhelming force of 181 ships and between 30,000 and 40,000 men to invade Malta, then not as fortified as it later became – the magnificent walled city of Valetta was built by the knights as a response to their victory in the siege and was not there in 1565. The Turks affected a landing easily enough and, with their vast superiority in numbers – there were under 700 knights and 9000 followers – they should have won. However, the knights were obdurate, the Turks expended vast resources in finally capturing a small structure, Fort St Elmo, rather than simply isolating it, and the invaders were both ill-led, in terms of generalship, and badly affected by disease. The Knights held out in their other forts until a relief squadron could arrive from Sicily and the Turks withdrew, never again to express their power so far west. This rare example of an island victory does not defeat the thesis, the Turks should have won and the island did need relief from outside to finally beat off the attackers (see Bradford, 1961, for a fine account of the siege).

Some islands have been taken by almost every passing power. Cyprus was colonised by Myceneans and Archaeans (fourteenth to eleventh centuries BC). It was settled by Phoenicians (ninth century BC); came under Assyrian influence (seventh century BC), was Greek (fourth century BC), Egyptian (Ptolomeic) (third century BC) and Roman (58 BC–AD 395). When the empire split, Cyprus became Byzantine (this power had to fight for the island against the Muslims from the seventh to tenth centuries). The island was taken by Richard the Lionheart of England (1191) who sold it to a French noble, Guy de Lusignan, whose family kept the island for three centuries before the Venetians took it (1498). The Turks conquered Cyprus next (1571), later ceding it to British administration (1878). The British formally annexed Cyprus (1914) before offering it to Greece in 1915 if that country would join in World War I. The Greeks refused and the island remained British until its independence in 1960. Even then as an independent island, Cyprus had guarantor powers, Britain, Turkey and Greece, who were to safeguard the new state's sovereignty. However Britain and Greece were unable to react in time to stop the island's latest invasion, by Turkey again, in 1974, since when the island has been divided, de facto, into two separate states.

The colonial era is littered with cases of islands being invaded and re-invaded by European powers. A Caribbean example is St Lucia, which was seven times French and seven times British before independence; its people speak a French patois and drive on the left. In the Indian Ocean, Mauritius was Dutch from 1598 (it was uninhabited prior to what was, in this case, true European discovery), French from 1721 to 1814 when it was ceded to the British at the Treaty of Paris. The British had been initially defeated off Mauritius in 1810 at the naval battle of Grand Port, the only naval victory the French were able to chisel in the stone of the Arc de Triomphe, but returned the same year and conquered. A Pacific case is the Marshall Islands, which were Spanish, German, Japanese and American.

Nor have the invasions stopped. Since the 1970s there have been outside incursions into East Timor (the Indonesian invasion in 1974 and Australian-led peacekeepers in 1999). Turkey invaded Cyprus in 1974. The Comoros Islands had three coups under French mercenary Robert Denard (real name Gilbert Bouregaud) in 1975, 1978 and 1989. During the 1978 and 1989 coups the Presidents were killed. French paratroops imposed order in 1989. In 1995, then aged 66, Denard invaded again and France restored order, again. The Seychelles suffered a failed mercenary coup in 1981. The Falkland Islands had the Argentinian invasion and British counter-invasion in 1982; Grenada, the American invasion in 1984. Sri Lanka had Indian 'peacekeepers' from 1987–90; the Maldives a failed mercenary coup in 1988, which was put down within hours by India. Haiti had American troops to support democracy from 1994–6. Anjouan was invaded by troops from elsewhere in the Comoros Islands in 1997. Libya has shelled Lampedusa (Italy); and a number of islands in the Persian Gulf, especially the Iranian Kharg Island with its oil terminal, were caught up in the Iran–Iraq war in the 1980s. Shortly after Grenada had been invaded in 1984, the Commonwealth advised its smaller members to consider regional security arrangements or defence links with larger regional powers to enhance their security. It also recommended that a discrete posture in foreign affairs might be advisable. 'Small states to stay weak and vulnerable' was one newspaper headline (*Independent*, 2 October 1985; see Houbert, 1992, for a discussion of security issues amongst the islands of the Indian Ocean; also Lemon, 1993).

Powerlessness is not normally associated with military battles, it is a daily dependence on more powerful forces off the island to take decisions that affect the life and livelihoods of islanders. If an island is an offshore part of a larger political unit it is inevitable that it will not be at centre stage when decisions are taken. Its needs for, say, infrastructural investment to help a small population will have to compete against expenditure needs of greater populations elsewhere. Some islands have tried to scale up their political and economic influence, this is a theme that will be treated in Chapter 8. For independent islands the powerlessness certainly remains in military terms, but the more usual problem is being at the whim of outside forces that control the economy. Thus, those many Pacific Island nations whose chief export is copra – coconut oil – and other coconut products are at the mercy of the world market price. In the late 1990s this was low and Kiribati, the Marshall Islands and others were in difficulties as a result. A similar story could be told of Caribbean, Indian Ocean islands and some in the Pacific, such as Fiji, dependent upon sugar production. The 'all the eggs in one basket' scenario, inevitably leads to powerlessness in the face of world market prices, and no island can hope to manipulate these prices as each island's absolute contribution to world supply is low.

Resources

Small islands often suffer from a restricted range of resources. There may well be significant opportunities for money making but, typically, from only one or a small number of products. Thus Nauru in the Pacific is a small phosphatic knoll and if, as has happened, the very being of the island is stripped away for the production of fertiliser, a good deal of money is generated as a result. Thus the Nauruans have been one of the wealthier population groups in the world, if at the cost of the devastation of its interior, which is often likened to the surface of the moon. Most Nauruans choose to live elsewhere. Their island's resource enables them to do so. So islands do not have to be poor, though many are; what is common is that their range of resources will be limited – in Nauru's case one of the restricted number of resources just so happened to be extremely valuable.

More typical is the situation in which islanders struggle to make a living from their restricted resource base. The limitation is simply a matter of scale and location. The small area of islands means that their terrestrial resource base must perforce be limited to that of its region. Large continental countries such as the USA can produce everything from winter wheat through cotton to wine as they encompass several bio-climatic zones. An island typically sits in just one. Some, such as Tenerife in the Canaries, are mountainous enough to experience a variety of habitats as the climate changes with height, but the small scale comes into play and the absolute size of the opportunities provided by the different habitats will be small. Most islands do not have the vertical zonation of Tenerife and so their terrestrial resource base is perforce even more limited.

Small island geology is often relatively restricted: in simple terms Mallorca is limestone, Hawaii is basaltic, the Maldives are coral. There are more complex scenarios, but usually islands again are granted just whatever few different mineral resource opportunities their restricted geology has provided. The resources might be valuable – Brunei, a small part of the large island of Borneo, has oil – but the range is again limited. Utilising the restricted range of resources in an insular setting usually sees the adoption of one of two strategies, as will now be discussed.

All the eggs in one basket

One way of coping with island resource restrictions is to maximise production of the resource that provides the best return. This strategy inevitably hits up against the matter of scale. For a small island to obtain sufficient volume of product to compete effectively in the market place it may well have to turn its entire or nearly entire productive capacity over to one good. This is the equivalent of the proverb about putting all your eggs in one basket: should the basket fall (should demand or, alternatively, supply of the one product fail), you have lost all your eggs (you have said goodbye to your economy). There are many examples of this. King and Young (1979) wrote about the Aeolian Islands in the Mediterranean where the productive focus was upon wine production. The vines in the nineteenth century became infected by *phylloxera*. The industry was lost and islanders were forced to leave. McElroy and de Albuqueque (1990) link the opportunities provided by an island's focused production doing well and then the devastation of its downturn with population dynamics, and demonstrate that the population fluctuates in association with the economy.

Resource use may also change with time as demands for different goods and services vary. Thus on the Falkland Islands, once the British had established their rule in 1833, they had to find an economy for their new colony. Initially the islands were used as a base for the refitting and resupply of vessels on the Cape Horn route. By around 1870 the interior of the islands had begun to be opened up to agriculture. The Falklands economy became focused on the production of wool from a number of company-owned estates (Royle, 1985). By the 1970s that economy and the society to which it gave rise had become rather moribund. Then in 1982 the Argentinians, largely for domestic political reasons, invaded. The British, under Margaret Thatcher, responded vigorously, expelled the Argentine forces and retook the islands. The Falklands were then centre stage, the British did not meet universal acclamation for the way in which this rather nineteenth-century colonial war had been conducted, and had to be seen now to develop the colony upon which so much blood and treasure had been spent. Agriculture continued to be modernised, there were new opportunities too for service industries and then another resource around which the newly prosperous Falklands economy could become focused conveniently arose. This is the sale of fish licences. There are controls to try and prevent overfishing, but the islands are well aware of the 'all the eggs in one basket' scenario. In the late 1990s oil was being sought, although initial trial drilling proved to be disappointing. In sum, the Falkland Islands have had a series of economies, in each case depending on one particular activity (see Chapter 8).

Occupational pluralism

A second strategy is occupational pluralism, which will be discussed in detail in Chapter 4. This is where islanders engage in a range of activities to tap into whatever economic opportunities are available. In a subsistence setting this may see them doing a little farming, some fishing, perhaps finding labouring or service opportunities when available, perhaps some craft work for tourists. Individual occupational pluralism and its modern derivation of multi-functional business operations still takes place. Thus on small islands one often finds commercial activities dominated by one or two horizontally integrated companies. On Majuro Atoll in the Marshall Islands, a company called Gibson's runs a cinema, a downtown supermarket with other smaller shops elsewhere, a restaurant, a garage, a car-hire business with outlets at the airport and in town, a wholesale supply company, a hardware store and a shipping business. Gibson's rival is Robert Reimers Enterprises with a hotel, hairdressers, supermarket and other stores, a shipping business, a water company and a mariculture operation. Major amounts of money cannot be made from any one business in the island setting; the scale of the demand is limited, and Majuro has only 30,000 people, the entire Marshall Islands only about 63,000. So the entrepreneurs maximise their returns by horizontal linkage, by using their time and talents to run what is a series of small businesses, not being able to run one large one.

Resources and the environment

Islands may have not just a restricted range of resources on which to base an economy, but also a restricted range and quantity on which to sustain life. The most common problem is water supply. Small islands, as small spaces, receive absolutely little precipitation. On islands throughout the world, one can observe measures taken to maximise the utility of precipitation, measures that are not necessarily needed in mainland situations. Bermuda uses its roofs as water catchments; the stepped limestone block roof with its systems for guiding water into cisterns is one of the most notable features of Bermudan vernacular architecture. Such supplies can be supplemented by groundwater, any streams and by whatever reservoirs can be built. Thus on Mallorca, there are reservoirs in the Sierra del Norte and, for generations, the farmers have abstracted water from the aquifer to irrigate their crops. However, the history of irrigation in Mallorca says something about the fragility of water supplies in the island setting. Traditionally farmers used wells and a system called the *noria*. This was a donkey-powered, low-tech arrangement where the donkey would spend its life walking in a circle attached to a beam which would turn a post horizontally. On the post were a series of cogs that, through basic gearing, would drive a vertical wheel that dipped down into the water. Attached to the wheel were terracotta pots, which would bring up water, which, after reaching the top of the wheel's rotation, would spill their contents into a trough. However, over-abstraction and/or technological development saw the donkeys replaced by wind pumps, large, elaborate affairs that still dot the rural landscape of the island, though now no longer in use. These could produce more power than a donkey, drive further into the aquifer, allowing more abstraction. In time the windpumps were replaced by the present electrical or diesel powered pumps and again extraction rates could increase. The situation now is that there has been so much taken from the underground water supply, because of a mixture of increasing demand and the technical ability to pump from greater depths, that the aquifer is increasingly penetrated by sea water and some of the pumped water resources are now brackish, especially in the summer. Mallorca's problems are compounded by the fact that the peak demand comes in the hot summers from both agriculture and from the island's huge tourist trade, whilst the Mediterranean climate of the area delivers almost all supplies of precipitation in the winter. In recent years such have been the problems that Mallorca has on occasion had to import water in tankers from mainland Spain. In the late 1990s, a new – and expensive – water desalination plant had to be opened but its capacity is already seen to be not enough.

Water problems, as with so many other insular difficulties, are maximised in the cases of atolls, where natural and man-made storage facilities are very limited and there is an immediacy in supply and consumption of water that is found in few other geographical settings. The larger of the discontinuous land area of atolls, those patches where the thin, snake-like island fans out, may well sustain a sub-surface freshwater lens. This can support wells and limited pumping, though there is the constant concern that over-extraction may lead to, as at the much larger scale of Mallorca, sea water penetration and increasing salinity. Thus a drought on an atoll cannot see the islanders simply turn to greater abstraction from lenses, except as a short-term emergency measure. On atolls there is usually arrangements for water catchments. On Majuro Atoll, many of the houses and institutional buildings have aluminium roofs and these form, as in Bermuda, an important catchment for domestic supplies. There is also a reservoir filled with water caught from the airport runway. On Majuro the airport is basically a concreted section of the atoll. The runway is sloped so that water drains towards the reservoir next to the airport. There is an immediate use made of it as there is just no long-term storage capability. Thus if the supply from the airport surface is interrupted, in a short space of time the atoll can be in grave difficulties. The central Pacific region suffered a drought in early 1998 as a result of the El Niño-inspired weather changes. Within a couple of months of receiving less than average rainfall, the situation on Majuro and the rest of the Marshall Islands was bad enough for the government to declare a state of emergency. Japan stepped in with three small reverse osmosis plants for desalinating seawater and then the Americans, under the Compact of Free Association, which gives them relict rights and responsibilities for the Marshall Islands, declared them to be a Federal Disaster Area. This released 75 per cent funding for a series of huge reverse osmosis plants, brought in by the world's largest plane, the arrival of which saw most of Majuro's population gather at the airport. Most of machinery was placed at Majuro Airport to abstract water from the ocean and put it straight through the reverse osmosis process into the reservoir. Before the reverse osmosis plants were put in operation, there had needed to be increased abstraction from the lens under Laura at the western end of the atoll, to 250,000 gallons per day. This was to the discomfiture of islanders at Laura who were forbidden their usual access to it for the greater good of the majority of Majuro people who live at the eastern end of the atoll and were without their water supplies from the airport. Meanwhile, water was cut off for long periods everywhere and people queued up for drinking water from the Japanese-provided small reverse osmosis operations (Figure 3.4). Other plant from the Americans was taken to Jaluit Atoll and to Ebaye, the Marshall's other urbanised island at Kwajelein Atoll, where many Marshallese work at the

Figure 3.4 People queuing for desalinated water, Majuro, Marshall Islands, 1998

American base on Kwajelein Island. Prior to that many residents of Ebaye had to take the boat to Kwajelein to fill jugs with water made available by the American forces as a goodwill gesture. On Kwajelein, the Americans were making maximum use of non-potable treated water, had increased abstraction from the lens until the water turned brackish and they had to engage stand-by desalination plants. The Americans had also provided a new water catchment in the north of Ebaye, but a catchment is no good without precipitation.

In short, this was a genuine crisis in the Marshall Islands, the low point saw the reservoir at Majuro airport down to only 1.4 m gallons, a few days' supply at most and less than the 2 m gallons needed to pressurise the system and get water flowing down the pipes to the town. This crisis was caused by the atoll being almost totally reliant upon rainwater, and thus it emerged inside a very short time period. The American president announced the declaration of the state of emergency in the third week of March and said in his announcement that the drought had started on 17 January 1998 (*Marshall Islands Journal*, 27 March 1998). This shows the problem of domestic resources in the small island setting.

This chapter has introduced some of the major issues associated with insularity, from the impacts of isolation and peripherality, through powerlessness to resource issues. It concludes the introductory section. The following chapters are thematic, starting with a consideration of islands in the historical period.

References

Key readings

Hintjens and Newitt's book on small tropical islands is comprehensive. The issue of scale can be followed up more generally in Ólafsson's work. King and Connell's recent edited book presents a range of accounts of migration in the island world. Bradford's history of Malta in the Great Siege is a fascinating tale of that rare event, an island beating off a determined aggressor. Finally, Marcus Clarke's novel can be recommended for the power of its depiction of the bleakness of the concept and, presumably, the actuality of the island as prison.

Bailey, E.E. (1983) *Republic of Kiribati: report on the 1978 Census of Population and Housing, Volume III*, Ministry of Home Affairs and Decentralisation: Bairiki, Tarawa, Kiribati.

Baum, T. (1997) 'The fascination of islands: a tourist perspective', in D.G. Lockhart and D. Drakakis-Smith (eds) *Island tourism: trends and prospects*, Pinter: London, pp. 21–36.

Beller, W.S. (ed.) (1986) *Proceedings of the interoceanic workshop on sustainable development and environmental management of small islands* (US Department of State: Washington).

Bendure, G. and Friary, N. (1995) *Micronesia*, Lonely Planet: Hawthorn, Australia, third edition.

Brand, I. (n.d.) *Penal peninsula: Port Arthur and its outstations, 1827–1898*, Regal Publications: Launceston.

Bradford, E. (1961) *The Great Siege: Malta 1565*, Hodder and Stoughton: London.

Clarke, M. (1875) *His natural life*, R. Bentley: London (republished in 1995 by The Tasmanian Book Company: Launceston).

Court, Y. (1987) 'Planning urban conservation: the case of Ribe, Denmark', in R.C. Riley (ed.) *Urban conservation: international contrasts*, Dept. of Geography Portsmouth Polytechnic, Occasional Paper, 7.

Dodds, K. (1988) 'Unfinished business in the South Atlantic: the Falklands/Malvinas in the late 1990s', *Political Geography*, 17, 6, pp. 623–6.

Dolman, A.J. (1985) 'Paradise lost? The past performance and future prospects of small island developing countries', in E. Dommen and P. Hein (eds) *States, microstates and islands*, Croom Helm: London, pp. 70–118.

Eriksen, T.H. (1993) 'Do cultural islands exist?', *Social Anthropology*, 1.

Ferguson, R.J. (1986) *Rottnest Island: history and architecture*, University of Western Australia Press: Nedlands.

Hamilton-Jones, D. (1992) 'Problems of inter-island shipping in archipelagic small-island countries: Fiji and the Cook Islands', in H.M. Hintjens and M.D.D. Newitt (eds) (1992) *The political economy of small tropical islands: the importance of being small*, University of Exeter Press: Exeter, pp. 200–22.

Hintjens, H.M. (1992) 'France's love children: the French Overseas Territories', in H.M. Hintjens and M.D.D. Newitt (eds) (1992) *The political economy of small tropical islands: the importance of being small*, University of Exeter Press: Exeter, pp. 64–75.

Hintjens, H.M. and Newitt, M.D.D. (eds) (1992) *The political economy of small tropical islands: the importance of being small*, University of Exeter Press: Exeter.

Houbert, J. (1992) 'The Mascareignes, the Seychelles and the Chagos, Islands with a French connection: security in a decolonised Indian ocean', in H.M. Hintjens and M.D.D. Newitt (eds) *The political economy of small tropical islands: the importance of being small*, University of Exeter Press: Exeter, pp. 93–111.

King, R. (1993) 'The geographical fascination of islands', in D.G. Lockhart, D. Drakakis-Smith and J. Schembri (eds) *The development process in small island states*, Routledge: London, pp. 13–37.

King, R. and Connell, J. (1999) *Small worlds, global lives: islands and migration*, Pinter: London.

King, R. and Young, S. (1979) 'The Aeolian Islands: birth and death of a human landscape', *Erdkunde*, 33, pp. 193–204.

Lemon, A. (1993) 'Political and security issues of small island states', in D.G. Lockhart, D. Drakakis-Smith and J. Schembri (eds) *The development process in small island states*, Routledge: London, pp. 38–56.

Lynn, M. (1984) 'Commerce, Christianity and the origins of the "Creoles" of Fernando Po', *Journal of African History*, 25, pp. 257–78.

Lynn, M. (1990) 'Britain's West African policy and the island of Fernando Po, 1821–1843', *Journal of Imperial and Commonwealth History*, 18, 2, pp. 191–207.

McElroy, J.L. and de Albuqueque, K. (1990) Managing small-island sustainability: towards a systems design, *Nature and Resources*, 26, 2, pp. 23–9.

Milton, G. (1999) *Nathaniel's nutmeg: how one man's courage changed the course of history*, Hodder and Stoughton: London.

O'Carrigan, C. (1995) 'Cockatoo Island: an island of incarceration in Sydney harbour', in J. Pearn and P. Carter (eds) *Islands of incarceration: convict and quarantine islands of the Australian coast*, Amphian Press: Brisbane, pp. 61–78.

Ólafsson, B.J. (1998) *Small states in the global system: analysis and illustrations from the case of Iceland*, Ashgate: Aldershot.

Pearn, J. and Carter, P. (1995) *Islands of incarceration: convict and quarantine islands of the Australian coast*, Amphian Press: Brisbane.

Robinson, S. (2000) 'A whiff of revolt on Spice Islands', *Time*, 30 October 2000.

Royle, S.A. (1985) 'The Falkland Islands, 1833–1876: the establishment of a colony', *Geographical Journal*, 151, pp. 204–14.

Royle, S.A. (1994) 'Changes to the Falkland Islands since the conflict of 1982', *Geography*, 79, 2, pp. 172–6.

Usher, R. (1996) 'A Pacific ground zero: France's decision to test H-bombs stirs outrage', *Time*, 26 June 1996.

Young, D. (1996) *Making crime pay: the evolution of convict tourism in Tasmania*, Tasmanian Historical Research Association: Sandy Bay.

4 Islands in the past

It is the Millennium in Kiribati, too

The world is still some way from being a global village. Lifestyles, opportunities and access to the media and material goods continue to vary. Mali is not Canada; most citizens of Mali do not have the same chances for obtaining wealth and the trappings of development as most citizens of Canada. The same is true of the island world. Take access to the media as an example. On islands in the developed world many people would read both morning and evening newspapers and their Sunday paper requires the sacrifice of a small, hopefully sustainably managed, forest. On their televisions, connected to cable or satellite dishes, there might well be in excess of 100 stations, including several 24-hour news channels. There is the Internet with all its facilities for news, comment and information; there is teletext. Their radios access dozens of stations. In short, there is far more information available to such islanders about the goings on of the world than they can possibly handle.

By contrast, take Kiribati. There is one radio station only, Radio Kiribati. It transmits for just a few hours per day, in a mixture of English and I-Kiribati, mainly local music with occasionally some very old western pop music. There are five-minute news summaries from Radio Australia and shorter local bulletins in English and I-Kiribati. The service closes at 9.30 p.m. and has breaks in transmission during the day. There is no television service. There is just one newspaper, *Uekera*, a weekly of 12 A4 pages. International newspapers, with the exception of the weekly *Marshall Islands Journal*, are not normally available. Visitors to Kiribati, and the I-Kiribati themselves have to wait for mere snippets of news of the world. The world is not a global village.

However, nor is Kiribati or any other developing world island nation stuck in some pre-modern time warp, forced to do things in the old ways. Modern technology and modern management systems exist here as anywhere else, it is just that some aspects of the western materialism are not as widely available. The Kiribati Broadcasting and Publications Agency is perfectly aware of what television is; it is just that there is not the money, either for the agency to start broadcasting or for many of the I-Kiribati to afford sets. There are videos and video rental shops so I-Kiribati are not cut off entirely from modern technology or Hollywood films; videos are used for education, too. Nor are the I-Kiribati, even those on the outer-islands, divorced from the modern world's system of government. Consider this quotation arising from fieldwork carried out on Tamana Atoll between 1971 and 1974:

> On all islands studied there appear to be two basic conceptions of development: innovations that come from government, and a greater opportunity to pursue Gilbertese ends [this was pre-independence, when these were the Gilbert Islands, part of the British Gilbert and Ellice Islands Colony]. Since there was no expectation of growth in the traditional society, the question of innovation did not arise. The idea of development is thus wholly associated with 'Tarawa' (synonymous with government) [South Tarawa is the nation's urban area and capital], what government does and institutes from outside. Most directly it is seen in the visible results in the form of airfields, roads, shipping services, medical dispensaries and services, Island Council Schools, coconut replanting and improvement services and the like. Since such infrastructure clearly leads to economic or social development, this view of development is sound enough.
>
> (Geddes *et al.*, 1982, p. 11)

So what one cannot do in this chapter, which looks at islands in the pre-modern age, is to associate historical developed world islands with today's developing world islands. Further, one cannot use modern material to try to identify and analyse some 'pre-industrial' society and economy on islands in the Third World. Kiribati may not have access to dozens of television channels but it is the Millennium period here as everywhere else. So what this chapter does is to take historical material, in the context of that past, where development did not come from Tarawa or any other synonym of government; development just did not come and most islanders had no expectation that things were going to get better. Island societies today, including Tamana, a remote part of Kiribati 'one of the most limited resource bases for human existence in the Pacific' (Geddes *et al.*, 1982, p. 1), are helped by aid, welfarism, some aspects of modern technology, including telecommunications (if not television), also, in this case, remittances. The impact of these phenomena will be considered in Chapter 8; here we consider the pre-modern era when they did not exist.

The chapter is not developmental; it does not take islands through any sequence of pre-modern to modern, pre-industrial to industrial. Nor is there any historical sequence to the structure of the chapter; it is just that most of the situations described pre-date the twentieth century. The chapter simply takes a thematic approach, which demonstrates how islanders and islands fared in eras that required them to be dependent upon domestic resources in a way that is not quite so necessary in today's globalised world. This is not to say that islands in the past were necessarily cut off from the rest of the world, as the first three sections below will illustrate. Even so, such interactions did not always do much to alleviate the population–resource balance problem that characterised many islands before the twentieth century. The chapter concludes with a consideration of this issue. Detailed examples in this chapter come from a range of places throughout the world and have been selected because of the level of information available.

Island connections

Even in the past, islands were often considerably involved with the world in terms of trade and participation in events. For example, the Shetland Islands were incorporated into the trading network of Europe (Smith, 1984). Thus a Hanseatic League trading booth is still evident on Whalsay. In the mid-eighteenth century

> the coast towns of Shetland [were] enriched by this concourse of foreigners, who go continually on shore both to buy and sell. . . . 'Tis this concourse of foreigners that makes all the trade of Shetland . . . they set up booths on the shore as in a fair, in which they sell wines, brandy and spices; and receive in return beer, bread, flesh and plants.
>
> (*London Magazine*, 1752, quoted in O'Dell, 1939, p. 308)

In the past, when land transport could be, and was, both difficult and dangerous, the sea was a highway, not a barrier, and some islands could be viewed as the equivalent of service stations on the modern motorway. Thus, centuries before the Hanseatic League, to the Vikings coming to Britain in the Dark Ages, some of the offshore islands were better known and better valued, at least initially, than the mainland areas. Note how many of the smaller British isles have an 'ey' or 'ay' placename ending, from the Danish word for island (Table 4.1). This is similar to the Faroes where, of the 18 islands, only Vagar and Mikenes do not have 'oy' at the end of their name (Eysturoy, Sandoy, Stremoy, etc.). Some of these islands went on to make good use of their sea connections. Witness Guernsey, where a book on St Peter Ports from 1680 to 1830 was subtitled '*the history of an international entrepôt*' when Guernsey's capital and chief port traded with Russia, Sweden, Denmark, Holland, Ireland, France, Portugal, Spain and the West Indies (Cox, 1999). A book about another trading island, Mallorca, used the word 'emporium' to summarise its historical economy (Donald and Abulafia, 1994).

Even the remotest islands were not all cut off from the world. Thus on St Kilda, the most distant archipelago of the Western Isles/Outer Hebrides of Scotland, islanders were not left to their own devices. St Kilda's landlord family, the MacLeods, sent their agent and a large retinue on annual 'predatory' visits from at least as early as 1549 to extract dues (Fleming, 1999, p. 183). Dozens of the landlord's men had to be supported by the islanders for about two months per year on this visit, another benefit for the MacLeods. In return the landlord replaced boats when necessary, and also helped in any emergencies. For example, after decimation by smallpox in 1727, St Kilda was repopulated to keep its economy and society viable. St Kilda was, thus, an integral unit of the MacLeod chiefdom and was not apart from the world.

Table 4.1 Scottish, Welsh and English islands with an 'ey' or 'ay' placename

Shetland	Scalpay
Bressay	Soay
Whalsay	Stuley
Orkney	Taransay
Burray	Vallay
Copinsay	Vatersay
Eday	Wiay
Egilsay	*Hebrides, Inner*
Gairsay	Colonsay
North Ronaldsay	Islay
Papa Sanday	Raasay
Papa Westray	Sanday
Rousay	Scalpay
Sanday	Soay
Shapinsay	*Welsh Islands*
South Ronaldsay	Anglesey
Westray	Bardsey
Hebrides, Outer (Western Isles)	Caldy
Berneray (twice)	Ramsey
Boreray (twice)	*English Islands*
Eriskay	Lundy
Grimsay	Walney
Hellisay	*Channel Islands*
Mingulay	Alderney
Pabbay (twice)	Jersey
Ronay	Guernsey
Sandray	

However, on historical St Kilda, the dues and the food for the MacLeod party as well as the subsistence needs of the local islanders had to be extracted from the unpromising resource base of the archipelago. There was little outside aid, no remittances, no welfare payments and no income from tourists to help.

If domestic resources failed an island simply lost its utility to the outside world. Take the case of Rutland Island, off the west of Ireland. Rutland was central to the development plan for William Burton Conyngham's County Donegal estate in the eighteenth century (advertised in *Faulkener's Dublin Journal*, 12–15 March, 1785). The key to the plan was to establish a fishing station, and to this end Rutland Island was to be laid out completely into a series of grid-plan streets. The main post office for the region was on the island. The plan depended on a prosperous fishery and in the best years, 1784–5, about £40,000 per annum was made and c. 1200 fishermen were employed. However in 1793 the herring deserted the coast. Rutland Island had insufficient other resources to maintain its modernity and with only one of the planned streets actually built, Conyngham turned his back on the island. Instead he devoted his attention to the mainland, principally the town of Burtonport which he founded in 1805. Rutland Island is now abandoned.

Another Irish case is in the seventeenth century when the British used

Inishbofin, County Galway to command the west coast and so built on this island the infrastructure necessary – harbour, fortress, accommodation, etc. – making Inishbofin, like Rutland, more advanced than the mainland opposite. Mention of Inishbofin, chosen by the British as a base at least partly because it could be defended, leads into the next section which considers the past use of islands as defensible sites.

Islands as defensive sites and stepping stones

In some ways, islands are easier to defend against minor attacks than mainland areas. The sea provides some protection against all but determined invaders, although if the islanders or their allies lose control of the seaways then the island is in difficulty indeed. Thus islands were often used as defensible sites. Sometimes the defence was against potential attack or nuisance from a neighbouring landmass. As well as Inishbofin, the British extensively fortified Alderney in the Channel Islands to discourage any French military excursions against England. Copenhagen is protected to its east by island fortifications. Halifax in eastern Canada was a good site for a British base in the mid-eighteenth century, not just because of its sheltered anchorage, but because its approaches could be guarded from McNab's Island. There are many other examples.

Islands could also be used as forward bases, stepping stones to facilitate military activity and perhaps colonial expansionism. For example, in the early years of colonial rivalries in the Spanish Main:

> the associates of the Puritan Earl of Warwick – whose Maritime interests embraced piracy, slaving and pirateering – used an island off Nicaragua as a base to attack Spanish trade, and, when expelled, briefly held Jamaica (1642–3).
>
> (Scammel, 1981, p. 471)

Somewhat later, the British used Bermuda as a base from which to attack America in the war of 1812, including the raid that saw the burning of the White House.

The island as stepping stone/defensible forward base has also worked in terms of settlement history in colonised areas. Manhattan Island's southern tip formed a good defensive site for the early Dutch settlement of New Amsterdam when Wall Street was not a financial area in the heart of the great city but marked the northern, landward fortification of the settlement. Such ventures were not always successful. In the 1580s the British colonisers on Roanoke Island, Virginia having lived for a couple of years largely from what they could trade with the local Indians, gave up and took passage home with Sir Francis Drake who visited in 1586. In 1587 another party settled at Roanoke but when they were looked for again in 1590, a delay occasioned by the alarums of the Spanish Armada, the party had vanished, whether through disease, massacre or amalgamation into local native tribes is not known.

Later and on the other coast of North America, early incursions into the northwest of the continent were channelled through Vancouver Island. Vancouver Island's city of Victoria is still the capital of British Columbia, not the larger, and otherwise more significant, later settlement of Vancouver on the mainland.

Off the east coast of Africa the island of Zanzibar, 'with its harbour and ships that tread the deep' (Stanley, 1872, p. 41) was for centuries a stepping stone for Arab and, later European incursions into East Africa. The Arabs had taken the island from the seventh century. It was used as an entrepôt, a convenient market place for trading the goods of the African continent:

> Zanzibar is the Bagdad, the Ispahan, the Stamboul, if you like, of East Africa. It is the great mart which invites the ivory traders from the African interior. To this market come the gum-copal, the hides, the onchilla, the timber, and the black slaves from Africa.
>
> (Stanley, 1872, p. 5)

Further:

> The greatest number of foreign vessels trading with this port are American, principally from New York and Salem. After the Americans come the Germans, then come the French and English. They arrive loaded with American sheeting, brandy, gunpowder, muskets, beads, English cottons, brass-wire, china-ware, and other notions and depart with ivory, gum-copal, cloves, hides, cowries, sesamum, pepper and cocoa-nut oil.... The Europeans and Americans residing in the town of Zanzibar are either Government officials, independent merchants, or agents for a few great mercantile houses in Europe and America.
>
> (Stanley, 1872, pp. 11–12)

From Zanzibar its rulers would prepare and mount raids and expeditions to the mainland for goods including ivory and slaves. Zanzibar also had a stepping stone role in some of the great European explorations of the 'Dark Continent'. Henry Morton Stanley made the above observations on Zanzibar when he was there to set up his journey to find David Livingstone. He took advice on how to organise his expedition from an Arab merchant with long experience of the ivory trade in the interior, one Sheikh Hashid:

> From the grey-bearded and venerable-looking Sheikh, I elicited more information about African currency, the mode of procedure, the quantity and quality of stuffs I required, than I had obtained from three months of study of books upon Central Africa; and from other merchants to whom the ancient Sheikh introduced me, I received most valuable suggestions and hints, which enabled me at last to organize an expedition.
>
> (Stanley, 1872, p. 23)

Note that the stepping stone facilitates travel *from* as well as *to* the mainland. To slaves from East Africa – 'black beauties from Uhiyow, Ugindo, Ugogo,

Unyamwezi and Galla' in Stanley's alarmingly politically-incorrect prose (1872, p. 5) – Zanzibar was a stepping stone out of Africa to whatever fate held in store for them elsewhere. On the other side of the continent, Goree Island off Senegal performed a similar role, as a secure place where slaves could be penned until transportation was available.

Other security uses for islands in the past have been as prisons, as mentioned in the last chapter, or as places to which people needing to be separated from normal society could be confined. This included the use of islands as lazarettos or leper colonies. A prime example was Spinalonga Island off Crete. This island was fortified by the Venetians in 1579 and here they managed to hold off the Ottoman Turks for 50 years longer than the rest of Crete. In 1904 the authorities decided to use the island as a leper colony and all the lepers in Crete were moved there, later joined by others from elsewhere in Greece until the island was abandoned in 1957. Off Australia, Peel Island, Queensland was used as a quarantine station as well as a lazaretto until its closure in 1959 (Ludlow, 1995).

Islands as chess pieces

In the past, the utility of islands because of a strategic location, some comparative advantage – such as an ability to produce a plantation crop – or simply to deny their use to a rival, would see islands contested for by outside powers. This process reached its peak in the colonial era. Islands are normally powerless in the face of superior outside forces and history is littered with incidents of islands being besieged, invaded and conquered (see Chapter 7). In regions of contestation, such as the Mediterranean – in classical as well as colonial times – the Caribbean and, on occasion, the Indian and Pacific Oceans, islands might become chess pieces – pawns – in the struggles for power and ascendancy between outside powers. This issue was introduced in Chapter 3, but deserves longer treatment here since being a chess piece was a shared experience of many islands in the historical period, and because this process further exemplifies the interaction of islands with the outside world in times past. We will consider an example of one island in detail here, a fine example of a Mediterranean pawn – the island of Crete.

The case of Crete

Crete is an exposed island, far to the south of the European landmass, about 100 km from continental Europe; 200 km from Asia (Anatolia) and 300 km from Africa. This location gives it a strategic comparative advantage, which is still utilised; Crete is an important American airbase and its massive Souda harbour is a NATO as well as a Greek facility. Its isolation makes it particularly vulnerable to outside attack, although its relatively large landmass (8261 km^2) and associated scale of population and economy, makes it a tough prize to take for any but the most prepared adversary. However, Crete's adversaries seemed often to have been sufficiently well prepared, as its violent history of invasion and counter-invasion demonstrates.

Crete's principal claim to fame actually rests on its endogenous early civilisation, the Minoans, whose period dates from about 3000 BC and whose best known remains are the palaces of central Crete, most notably Knossos. Crete even became something of a colonial power itself in the local region with outposts in Syria, Arabia and other islands in the region, including Sicily. But around 1450 BC, most Minoan palaces were destroyed by fire; Knossos survived for another 50 years and was actually embellished but, perhaps, by Mycaenians, not Minoans. Around 1400 BC, Knossos, too, was destroyed by pillage and fire and the Minoan era came to an end, although a few Minoans continued to live in seclusion in Eastern Crete. New masters came from the north, i.e. Greece, with the Dorians, their invasion coming around 1200 BC. The Dorians gave Crete a Greek culture and character that has survived through the many other invasions and changes of ownership which succeeded them. The Romans eventually conquered the island, although Crete was the last Greek region to become Roman. With the division of the Roman Empire in AD 395, Crete became Byzantine. Arabs took the island from the Byzantines in 824 before being expelled in their turn by a Byzantine invasion in 963. The Byzantines resettled the island that had been decimated during the Arab period. In 1204 the Fourth Crusade vanquished Byzantium and Crete was part of the booty. As an important staging post between East and West, it was the Venetians, a trading people, who purchased Greece, although for a couple of years the Genoans, having the same idea for Crete, actually took it, without authority, before the Venetians were finally able to establish their control in 1211. Crete was part of the Venetian Republic for over 400 years, although throughout much of that time the Cretans struggled against their unwelcome Catholic masters (the locals, then as now, being Orthodox Christians). However, the threat to Christianity generally from Islam, especially a revived Constantinople which fell to Ottoman rule in 1453, saw all the Christians on Crete band together. Crete, for a period, was a bastion of Greek culture, until in 1641, when the island fell once more under the sway of Islam with an invasion of the Ottoman Turks. As mentioned, Crete was a tough target for an island given its size and rugged terrain, and although the Turks took Chania in 1641, they were unable to complete their conquest until 1669 when they took Candia (modern Heraklion). The Turkish occupation was even more unpopular to the Cretans than the Venetian occupation and the island was under fairly continuous revolt from 1770 to 1898 as the Christians chafed under the discriminatory rule of the Turks.

Greeks from the mainland invaded in 1897, but were put off the island by European powers who maintained fleets in Souda Bay, for the Great Powers had long eyed Crete as a potentially attractive asset for their own chess games. In 1898 Turks in Heraklion murdered 14 British soldiers and a Vice-Consul, and the Great Powers expelled the Turks from Cyprus. The island was divided up with the Italians holding the southeast, the French the east, the British the centre and the Russians the west. The Great Powers placed the island under the rule of Prince George of Greece (actually himself Danish–German) as High

Commissioner. Prince George ruled under nominal Ottoman suzerainty with the protection of the four powers guaranteeing rights. The locals were not happy with Prince George's rule and, in 1906, he was replaced.

The local politician most prominent at this time was Eleftherios Venizelos and, in 1910, he became President of Crete before resigning to take over as Prime Minister of Greece that same year, where he worked to achieve Cretan enosis (unity) with Greece, which came in 1913. In 1922 there was a population exchange with Turkey with Greek refugees from Asia Minor being swapped with Islamic Cretans. This sealed Crete's Greek nature, but the island had still one last struggle to face when it became a pawn once more during World War II.

The Germans attacked in 1941. The first waves of troops were parachutists landing around Maleme airfield near Chania, where they suffered terrible casualties. The allied forces struggled ferociously but, as almost always with an island, the invaders, given sufficient *materiel*, were able to triumph. The Germans, with their air superiority, were able to supply and reinforce their troops and put 20,000 men on the island. After ten days Crete fell, with perhaps thousands of deaths on each side (Figure 4.1). Yet again, the Cretan locals struggled against this conqueror and guerrilla campaigns helped to tie up German troops until the island was liberated in 1945.

Table 4.2 sums up the situation regarding Crete as a chess piece. Usually the Cretans were revolting against whatever set of foreigners were ruling at the time, powerless and usually not-consulted islanders whose homeland was fought

Figure 4.1 World War II cemetery, Crete, Greece, 1998

Table 4.2 The rulers of Crete

Rulers	Dates
Minoans (indigenous)	3000 BC
Dorians	1200 BC
Romans	67 BC
Byzantines	325 AD
Arabs	824
Byzantines	963
Crusaders/Genoans	1204
Venetians	1211
Ottomans	1641 (final conquest 1669)
Great Powers partition	1898
Cretan independence	1910
Enosis with Greece	1913
Germans	1941
Port of Greece	1945

over, purchased and at one time parcelled out by and between more powerful forces from outside the island. Crete was a pawn indeed, and only one of many.

So far this historical chapter has had a political flavour, considering islands with regard to their interaction with the outside world. It has shown that in the past some islands were significant and well connected places and were often subject to contestation. Even islands much engaged in the world's activities were more likely to have had to rely completely or substantially on their own domestic resources than would be the case today, and so we will now go on to consider the issue of these resources.

The problems of resources and occupational pluralism

In historical times the difficulties of managing an island economy from domestic resources loomed large. Usually island life was a hand-to-mouth existence for most islanders, especially when there was little or no trade off the island. In such cases the 'all the eggs in one basket' scenario did not come into operation. Instead, islanders tended to be occupational pluralists, exploiting each of the limited opportunities provided by their insular realm in order to try to provide themselves with the range of foods and goods needed to sustain life. If the range of opportunities declined for some reason, islanders could be in trouble. The limiting case in this regard was Easter Island (Rapa Nui). This isolated Pacific island was settled by Polynesians, probably in the seventh century. At that time it was wooded and probably reasonably fertile. The population of the island may have reached 20,000 at one time, but before its rediscovery by other humans in 1722, the total had fallen to about 2000. The fall seems to have been a response to a reduction in the island's resources during the centuries of isolation, complete isolation once all the wood had been used up and the islanders could no longer make ocean-going boats, this largely denying them the use of the sea. It seems that the islanders were reduced to warring with each other over the remaining terrestrial resources; cannibalism was even practised. The environment was devastated and population tumbled as a result (Bellwood, 1979).

Rapa Nui is an extreme example of island population/resource problems. A more typical situation was that the islanders just resorted to scraping a living from any or all the opportunities provided. Occupational pluralism was characteristic, for example, of Cape Breton Island in eastern Canada in the nineteenth century. On Cape Breton there is little good land and early migrants occupied the best plots. When mass migration began to the island from western Scotland at the time of the Highland Clearances, there was little opportunity for the new people to support themselves fully from agriculture. Occupational pluralism resulted with islanders working the land, working woodlots and many were eventually also employed seasonally in the island's coal mines. In the 1830s:

> agriculture, generally speaking, [was] in a most slovenly and barbarous condition ... the farmers soon acquire the propensity ... of dabbling in pursuits unconnected with agriculture, such as fishing, hewing timber, building schooners, etc.
>
> (MacGregor, 1832, quoted in Tennyson, 1986, p. 117)

Cape Breton, a fairly small and poor island, could not cope with the Scottish migration and had to be subsumed into the mainland colony of Nova Scotia in 1820.

An unusually detailed investigation into occupational pluralism in a historic period can be made by considering the Irish Aran Islands in 1821.

The Aran Islands in 1821

Examining any historical society in depth depends on the collection in the past and survival to the present of suitable data. Fortunately there are some island groups with excellent sources that can be used, such as Iceland. The example used here is the Aran Islands, County Galway in western Ireland. The economy and society of this three-island archipelago can be examined in considerable detail given the very unusual survival of the enumeration material from the census of 1821. By 1821 Ireland had been part of the United Kingdom of Great Britain and Ireland for 20 years. However, the Irish census remained under separate control. Thus, whilst the British censuses of 1821 were counting heads and little else, the Irish authorities were engaged in a proper social survey. Sadly the 1821 census returns were amongst those classes of documents almost totally lost in the destruction of the Irish public records office in Dublin in 1921, when the building was caught up in the Irish Civil War. By chance and almost uniquely, the returns for the Aran Islands survived and they are analysed below.

The Aran Islands of Inishmore, Inishmaan and Inishere are three horizontally bedded slabs of carboniferous limestone. This is not the easiest geology on which to base a subsistence agricultural economy, for the rock is permeable and there is no running water. Fortunately there are, within the limestone, thin bands of clay and this leads to the appearance of springs and the traditional housing of the islands was positioned in relation to the location of places where water could be obtained. This is particularly clear on the central island, Inishmaan, where the housing, though in four clusters, is basically linear in pattern, aligned along the appearance of the clay where water was available. The people in the houses were basically peasants engaged in an almost entirely subsistence lifestyle. Their principal non-subsistence necessity was the requirement to garner sufficient cash to meet the rental payments on their land. This being Ireland, here as usually elsewhere, the land was owned by an absentee landlord, in this case a Reverend J.W. Digby who lived in Dublin.

Digby employed an agent resident on the islands. There was also a constable, a priest, a coastguard officer and three schoolteachers (one for each island). Except for the agent and two teachers, these middle-class outsiders all lived in Kilronan on Inishmore, the nearest thing Aran (the three islands together are called Aran) had to a town. These would have been the 'few protestants that are put there by the government to tyrannise' the Roman Catholic islanders in Liam O'Flaherty's barely disguised Aran novel, *Skerrett* (1932, p. 7). There were no service activities other than churches and one publican, and islanders had to travel to the mainland or rely on itinerant pedlars for anything that they could not themselves produce. The islanders made their own clothing, including shoes called *pampooties*, heel-less slippers manufactured from hide. They also made their own fishing equipment, including light boats – a local version of the Irish *curragh*, traditionally consisting of skins, later tarred canvas, stretched over a wooden frame. They built their own houses – from the limestone rock found everywhere on the islands – and roofed them with island-grown rye thatch. This was not an entirely cash-free or subsistence society, but it was clearly one where self-reliance and the abilities and skills to produce food and non-comestible goods from the restricted insular environment were necessary. Local production certainly included spirits, *poteen*; although, being carried out illegally, without payment of excise duty, this activity does not feature in the census returns. Nor does smuggling, though that surely went on as well.

Table 4.3 gives the basic population and density statistics for the islands in 1821. At the 1991 census the population totals were 836, 216, and 270 for Inishmore, Inishmaan and Inishere respectively and the far smaller populations were helped in their occupation by the paraphernalia of a western welfare state as well as modern industries such as tourism which is particularly significant on Inishmore. In 1821, things were not like that, as the census returns reveal.

Table 4.3 Population and density, Aran Islands, Ireland, 1821

Island	Population	Area km²	Population density per km²
Inishmore	2285	30.45	75
Inishmaan	387	9.1	42
Inishere	421	5.7	74

The considerable majority of Aran households in 1821 were nuclear families headed by a farmer, 296 of the total 493 households were of this nature. Here is where the evidence for occupational pluralism comes in, for only 96 of these people were recorded as being a farmer only. The rest had more than one occupation, three not being uncommon (Table 4.4). Farming was the most significant, they were always farmers first; the other two subsidiary occupations frequently added were fishing and kelpmaking, of which more below.

Islanders not in farmers' households or those headed by one of the few professional fishermen (all were men) were in particular difficulties in 1821, especially in some of the households headed by widows. If a widow had grown children who had taken up the land things might be alright; there were 25 land-holding three-generation households where the grandmother was widowed, and in 11 cases the widow was recorded as a farmer, including one woman whose age, improbably, was 100. Other widows and their families had no direct access to the produce of the land and, in a largely subsistence society, had to earn a living in other ways. Some widows were netmakers, servicing the fishing industry. Other women within nuclear families were sometimes described as dressmakers or had some other textile-related occupation, but they were probably producing clothing for the family circle, rather than as professionals.

Other disadvantaged families were those headed by landless labourers. As in every place and age, these were poor people and survived by taking whatever jobs might come up, at spring planting and harvest especially. Most tried to rent

Table 4.4 Landholding and occupational pluralism, Aran Islands, Ireland, 1821

Island	Holding size							
	More than 1 cannogara				*Less than 1 cannogara*			
	Farmer only		*Dual Occupation*		*Farmer only*		*Dual Occupation*	
	n	*%*	*n*	*%*	*n*	*%*	*n*	*%*
Inishmore	54	32.9	110	67.1	8	20	32	80
Inishmaan	18	43.9	23	56.1	1	20	4	80
Inishere	13	31.7	28	68.3	2	40	3	60
Aran Islands	85	34.6	161	65.4	11	22	39	78

Source: 1821 census enumerator's returns (Royle, 1983).

Notes
1. A cannogara was a local land measure. Its size varied from 7.1 to 9.2 ha in different parts of the Aran Islands.
2. The second occupation of farmers holding more than 1 cannogara was:
   Kelpmaker 38.7%
   Labourer 35.6%
   Fisherman/boatman 20.2%
   Crafts/services 5.5%.
3. The second occupation of farmers holding less than 1 cannogara was:
   Labourer 43.6%
   Fisherman 20.2%
   Kelpmaker 18%
   Crafts 10.3%.

small gardens to produce some of their own food, as did some of the professional fishermen of the village of Killeany on Inishmore.

Aran farming was largely subsistence. The principal product was potatoes,
which formed the bulk of the islanders' diet. Those with enough land and
capital might also keep a cow or two. This was principally for milk, but an
important by-product was calves, which thanks to the sweet, if sparse, year-long
grazing were sought after and graziers from the mainland might come out to
buy. Calf production for some, then, was one source of cash. It was mentioned
above that the Aran limestone geology was difficult; this was certainly the case
for those needing to water stock and not until much later did the innovation of
constructing small water catchments in the fields linked to troughs become
adopted (Klimm, 1936). In the nineteenth century, at times of drought, the
animals would have to be swum or ferried across to the mainland with its better
water resources.

Other difficulties were the limitations of the land. It was not simply that the
amount was small and finite; it was that only some of the area was productive, as
only some of the land had soil. Carboniferous limestone can naturally produce
bare limestone pavement and such is the case in Aran. These are exposed rock
surfaces with clints and grykes, basically cracks, scarring the outcrop. In the
cracks, windblown soil and seed can collect, hence the sweet but sparse grazing,
but there is not naturally enough soil in which to grow potatoes. Thus the Aran
islanders had to engage in 'landmaking'. This was a backbreaking activity,
telling evidence of the sparcity of the islands' resource base. First, crushed
stones would be placed within the cracks on the bare surface. Then a layer of
stones would be spread, then layers of sand from the beach, then seaweed for its
organic content, more sand, more seaweed, etc., until finally the field would be
finished with a layer of topsoil, often brought in from the mainland. The made
land produced a plaggen soil of reasonable fertility (Conry, 1971). The resultant fields were tiny, and to ensure the precious soil did not blow away, they
would be walled, the walls also conveniently consumed loose rock. This has
given Aran a landscape of tiny fields enclosed by thousands of kilometres of drystone walling (Figure 4.2). It is a distinctive and attractive landscape, but one
won at vast cost in terms of labour. Before marriage it was usual for a groom to
be able to indicate that he would be able to support his bride; for most islanders
that meant having access to sufficient made land. Rather than engage in landmaking or in addition to it, a man might wait to inherit land from his parents
before marrying. Thus grooms tended to be rather elderly, the lack of island
resources predicating social behaviour. One man in 1821, described as a
labourer, was at 55 still a bachelor, still living at home with his mother, she at
the improbable age of 100 being both the head of household and the *de jure*, if
not *de facto* farmer.

Fishing was a second occupation for many farmers. They would fish usually
from the *curragh*, crewed by two or three men. Aran had some full-time fishermen based at Killeaney on Inishmore who worked from larger sailboats called
hookers and traded on the mainland at Galway. However, for most, fishing, like

Figure 4.2 Drystone walling, Inishmore, Aran Islands, County Galway, Ireland, 1997

farming, was a second, seasonal activity recorded for farmers with smallholdings; only those with the largest holdings were not in need of further occupations, as demonstrated by the statistics in Table 4.4.

The third string to many Aran farmers' bows was kelpmaking. Kelp is seaweed. In the early nineteenth century, seaweed was used not just as fertiliser as it still is on the islands, but also as a chemical feedstock. The chemical, pharmaceutical really, was iodine and it was manufactured from partly processed material exported by the islanders. The weed would be washed onto the beaches after storms or people might wade out and cut it from rocks. Although normally taken off the beach, this did not mean that kelp was freely available, say, to the poor landless labourers, as rights to collect at certain spots were part and parcel of landholdings. So it was the farmers who took the kelp. It would then be dried, after which it would be burnt to produce a white residue, which was then broken up, and it was this seaweed ash which would be sold, the cash helping to meet rental payments. Each stage of the work was backbreaking. Aran residents had to work hard to wrest a living from their islands.

In sum, the Aran Islands in 1821 are a fine example of pre-industrial occupational pluralism. Individuals farmed, fished, and made kelp. The women made the clothing including the shoes; there were just two tailors on the islands. Islanders built their own boats and houses. They smuggled, they collected material washed up on the beaches, and, on occasion, they went sealing and took sea birds for their eggs and feathers. They turned bare limestone pavement into productive fields (Royle, 1983). The resources of their islands were stretched to the limit. No wonder then that when the year following the census just one of these resources collapsed, the potato crop failing thanks to adverse weather conditions, the islands faced grave difficulties. In fact they actually had to be helped from outside, in an almost modern relief process, although analysis of the attempts to relieve the 1822 Aran famine show the well meaning efforts to have been rather clumsily applied (Royle, 1984).

Sadly lessons were not learnt by the time of the much greater emergency of the Irish potato famine of the 1840s, when relief systems were both late to be fully applied and inadequate, and death and disaster ensued. The Great Famine was a case of the large island of Ireland having gone not for occupational plural-ism, but for the 'all the eggs in one basket' case. The Irish peasantry, especially in the west, had come to rely almost entirely on the potato for its subsistence. An exogenous disease from America, *phytophtera infestans*, struck; the potato was taken from the Irish poor, many of whom then became malnourished and died of disease – many more died in this way than actually of starvation. Others migrated, especially to Great Britain or North America; some survived on the derided and inadequate relief system (see the classic study of Woodham-Smith (1962) or the collected writings of Bourke (1993) on the topic). Actually, islanders on the small Irish islands tended to fare better than those in many parts of the mainland in the Great Famine. They had always been forced to practice occupational pluralism and could fall back on fishing and other resources; isolation also kept some of them free from the potato blight.

Primitive communalism

When island resources were stretched, the islanders would, on occasion, maxi-mise the utility to be made from them by employing them communally. Liam O'Flaherty, the novelist from Aran put it well, and also noted that the system was predicated upon need and was thus not necessarily stable.

> In the remote villages . . . I notice how fine and self-reliant the people are . . . they live as in a commune, all helping one another and hardly seeing a coin from one end of the year to the other. [Development would] make them give up their way of living and become cadgers.
>
> (1932, pp. 90–1)

On the Scottish St Kilda archipelago, in the one village which was on the island of Hirta, each morning the 'Island Parliament' would meet in the village street and the men decide what should be done that day. They farmed, they did a little fishing, but their principal activity, that which they were forced into because of the need for other sources of food and cash, was to exploit their archipelago's sea birds (Steel, 1972). Cash needed for the rent came from the sale of feathers and the birds were also a source of food and oil. Taking birds was a communal activity, with the men necessarily working closely together. Teams would be set up to lower agile men on ropes down the sea cliffs to gain access to the birds and eggs. The equipment was owned communally and the produce shared out amongst the island population, including those people too aged or infirm to participate directly in the food production process. Thus here, the restriction of island resources predicated the social organisation of this island society. St Kilda's communalism was not a planned or mannered decision, no contemporary Marx or Engels inspired the islanders to share their resources; aspects of communalism were adopted as simply a logical response to local, island conditions. St Kilda was evacuated in 1930.

Another historical example is that of Tristan da Cunha in the South Atlantic. Here organised settlement dates to the British annexation of the island in 1816, largely to prevent it falling into American or French hands, especially with Napoleon held in custody in St Helena, the next island north. Records show that the British occupation was poorly organised, that the garrison came close to starvation and the principal military event was the tragic wrecking of *H.M.S. Julia* on the island with the loss of 55 lives in 1817 (Royle, 2001). The British withdrew the garrison in 1817, but one of the marines, a Corporal William Glass, was permitted to remain behind as a resident and his family joined him. Two others, civilian stone masons called Nankevil and Burnell also stayed. From Glass's family and others who came to the island, often seafarers wrecked there, descended Tristan's present population of about 300, although the island's continued occupation has several times been called into question, especially after a longboat disaster in 1885 that drowned 15 of 18 able-bodied men (Royle, 1997). In the early days of Glass and the stone masons it was realised that they would have to co-operate together to wrest a living from their bleak, inhospitable home. Their communalism was formalised from the start and before the last of the naval party sailed away, Glass drew up the following remarkable contract, to regularise the operation of production on Tristan da Cunha, which was conceived by them as a firm, principally involved in sealing:

> We, the Undersigned, having entered into Co-Partnership on the Island of Tristan da Cunha, have voluntarily entered into the following agreement – Viz.
>
> 1st That the stock and stores of every description in possession of the Firm shall be considered as belonging equally to each –
>
> 2nd That whatever profit may arise from the concern shall be equally divided –
>
> 3rd All purchases to be paid for equally by each –
>
> 4th That in order to ensure the harmony of the Firm, No member shall assume any superiority whatever, but all to be considered as equal in every respect, each performing his proportion of the labour, if not prevented by sickness –
>
> 5th In case any of the members wish to leave the Island, a valuation of the property to be made by persons fixed upon, whose valuation is to be considered as final –
>
> 6th William Glass not to incur any additional expense on account of his wife and children –

The firm as such lasted only a few years. Its strict communalism could not survive the arrival of new residents who discovered that simply by staying on at Tristan da Cunha they could become partners in a sealing business. Nankevil's fleeing with some of the firm's money, and the loss of the firm's ship, both in 1823 at Cape Town did not help either. Burnell had earlier left the island. There was a revised constitution drawn up in 1821 but in time that lapsed, too, and Tristan society became less unusual as family, almost clan, units became established and agricultural activity was set up (Munch, 1974; Royle, 2000). However, traditions of equality and, more strongly, those of mutual help survived and survive. Thus, in the early twentieth century a visitor to Tristan, French explorer Raymond Raillier du Batty, wrote:

> As soon as we landed, I had noticed one man in the group with a certain air of distinction [Andrea Repetto, an Italian who had stayed on the island after being shipwrecked in 1892] … It seemed to me from the way in which the others treated him and from the precise manner in which he dealt with our proposed bartering that he must be in charge. I threw a glance round and asked aloud 'Who is the chief here?' He was the first to answer that there was no chief and that each man in the island was the equal of every other.
>
> (2000, p. 9)

The Tristan, St Kilda and other communal systems were a logical adaption of society's norms, given the sparcity and range of island resources. It was not the only possibility, however, and other small island societies looked instead to a strong chief to make hopefully wise decisions for the benefit of the island society. On some of the Irish islands one senior person was known as the 'King', an honorary position still held in some islands.

This chapter has shown that, in the past, islands might be useful to outside powers for their strategic value or for other attributes. Blood and treasure, on occasion, might be expended to secure them. For most islands, however, this was not the case, they were unimportant and insignificant places and their inhabitants struggled to wrest a living from them. As part of its consideration of island peoples and populations, Chapter 5 goes on to explain how the reaction of many islanders to their relative privations was to migrate.

References

Key readings

Bellwood's book is an important study of humankind and the Pacific. The book by Peter Munch on Tristan da Cunha emanated from a lifetime's fascination for that island. The work on the Aran Islands by Royle is a detailed survey of an historical island community. Tom Steel's book on St Kilda is a classic study of an island community which did not survive far into the modern period.

Du Batty, R.R. (2000) 'The voyage of the J.B. Charcot', *Tristan da Cunha Newsletter*, March 2000, pp. 9–10.

Bellwood, P.S. (1979) *Man's conquest of the Pacific*, Oxford University Press: Oxford.

Bourke, A. (1993) *'The visitation of God'?: the potato and the great Irish famine*, Lilliput Press: Dublin.

Conry, M.J. (1971) 'Irish plaggen soils, their distribution, origin and properties', *Journal of Soil Science*, 22, pp. 401–16.

Cox, G.S. (1999) *St Peter Port 1680–1830, the history of an international entrepôt*, Boydell and Brewer: Woodbridge.

Donald, J. and Abulafia, D. (1994) *Mediterranean emporium: the Catalan Kingdom of Mallorca*, Cambridge University Press: Cambridge.

Fleming, A. (1999) 'Human ecology and the early history of St Kilda, Scotland', *Journal of Historical Geography*, 25, 2, pp. 183–200.

Geddes, W.H., Chamber, A., Sewell, B., Lawrence, R. and Watters, R. (1982) *Atoll economy: social change in Kiribati and Tuvalu. Islands on the line team report*, Australian National University Development Studies Centre: Canberra.

Klimm, L. (1936) 'The rain tanks of Aran: a recent solution to an old problem', *Bulletin of the Geographical Society of Philadelphia*, 34, pp. 73–84.

Ludlow, P. (1995) 'Peel Island: quarantine as incarceration', in J. Pearn and P. Carter (eds) *Islands of incarceration*, Amphion: Brisbane.

MacGregor, J. (1832) *British America*, William Blackwood: Edinburgh.

Munch, P.A. (1974) *Crisis in Utopia*, Longman: London.

O'Dell, A.C. (1939) *The historical geography of the Shetland Islands*, Manson: Lerwick.

O'Flaherty, L. (1932) *Skerrett*, Wolfhound Press: Dublin (1982 reprint).

Royle, S.A. (1983) 'The economy and society of the Aran Islands County Galway in the early nineteenth century', *Irish Geography*, 16, pp. 36–54.

Royle, S.A. (1984) 'Irish famine relief in the early nineteenth century: the 1822 famine on the Aran Islands', *Irish Economic and Social History*, 11, pp. 44–59.

Royle, S.A. (1985) 'The Falkland Islands, 1833–1876: the establishment of a colony', *Geographical Journal*, 151, pp. 204–14.

Royle, S.A. (1997) 'The inside world: Tristan da Cunha in 1875', *Journal of Historical Geography*, 23, 1, pp. 16–28.

Royle, S.A. (2000) 'Historic communities: on a desert isle [Tristan da Cunha]', *Communities, Journal of Cooperative Living*, 105, 1999, pp. 16–20.

Royle, S.A. (2001) 'Perilous shipwreck, misery and unhappiness: the British military at Tristan da Cunha, 1816–1817', *Journal of Historical Geography*, forthcoming.

Scammell, G.V. (1981) *The world encompassed: the first European maritime empires, c. 800–1650*, Methuen: London.

Smith, H.D. (1984) *Shetland life and trade 1550–1914*, Donald: Edinburgh.

Stanley, H.M. (1872) *How I found Livingstone: travels, adventures and discoveries in Central Africa; including four months' residence with Dr Livingstone*, Sampson Low, Marston, Low and Searle: London.

Steel, T. (1972) *The life and death of St Kilda*, Fontana: Glasgow.

Tennyson, B. (1986) *Impressions of Cape Breton*, University College of Cape Breton Press: Sydney.

Woodham-Smith, C. (1962) *The great hunger: Ireland, 1845–9*, Hamish Hamilton: London.

5   Islands

People and migration

People migrate when forced to do so by irresistible circumstances at home, such as war or natural disaster, or when they perceive some betterment in social and/or economic circumstances to be gained from choosing to move. There are often high rates of emigration from small islands. It is not being on an island *per se* that leads to migration; people leave because of the circumstances they face and given the constraints imposed by insularity, adverse circumstances frequently arise in the small island situation. Small islands are often fragile places at risk from natural disaster; they are powerless places and have often been subject to military interference and warfare; they often have insufficient resources for their inhabitants to be able to make a decent living. These factors would all encourage migration. Further, small island society can be confining and restrictive because of its scale; this can also encourage the young to seek fresh pastures elsewhere.

By contrast, and characterising the recent past, there are now some islands where economic upturn has led to a return of out-migrants, although sometimes return migration is for retirement with money that has been made off-island. Further, the attractions of a pleasant island environment have seen holiday and retirement migration to some islands close to the vibrant economies of developed continents, Corfu is one good example (Lazaridis, Poyago-Theotoky and King, 1999). Thus, the migration experience of small islands today is varied but, for many, it remains one of out-migration. King and Connell's recent edited book on island migration has this in the introduction:

> Key themes which echo through many accounts are the restricted resource base due to such factors as smallness, rugged terrain and climatic vagaries; high rates of demographic increase; and the lure of more wealthy and modern economies elsewhere.
>
> (1999, p. 9)

To have people leave may not be totally negative to an island in that remittances from out-migrants can be an important source of income to those left behind. Migration can also relieve the pressure of population on resources. However, substantial out-migration from a small society can lead to difficulties of sustaining that society, and many small islands have gone on to lose their population entirely once the downward spiral of emigration has begun.

This chapter will consider the impact of both emigration and immigration to small islands. One resultant population issue of contemporary significance that will be discussed relates to the sharing of restricted spaces between different ethnic groups, such mixing often being the result of the immigration of labour supplies to islands, especially plantation islands, in the past.

Population and economic adjustment

Island economies are especially prone to boom–bust cycles as the 'all the eggs in one basket' strategy can have a severe downside if the product falls out of demand, becomes outdated by technological change, or, if a crop, becomes affected by disease. The reliance of the island of Ireland in the 1840s upon potatoes to feed its population, and the disastrous affect upon this population when an imported disease ruined the crop, is just the best-known example of these problems. The demographic response was dramatic: increased death rates, especially from disease, given the breakdown in normal sanitary practices and the proneness of malnourished bodies to infection anyway. Far more died from disease than actual starvation. Birth rates decreased as social relationships broke down and poorly fed people were less able and less willing to engage in sexual relationships and/or bear and care for children. There was migration to north-east Ireland, which was both less affected by famine and was industrialising, focused on Belfast; others migrated to Britain, to North America and other New World destinations. Kennedy *et al.*'s (1999) atlas on the famine years has a section on the 'unpeopling' of Ireland which demonstrates the spatial pattern-ing of the 20 per cent population decline from 1841 to 1851. (See also Mokyr (1983), for a detailed econometric analysis of the effects of the famine upon the Irish population.)

In an economic downturn, islanders can be driven from their homes. This has happened even in overcrowded Japan. Forty minutes by ferry from Nagasaki is the island of Takashima, which once supported 20,000 people through coal mining. The last mine closed in 1986 and within three years the population had dropped to 1500 and the smaller coal-mining island of Gukanjima had been abandoned entirely. Without the mines the dysbenefits of small island life were obviously more significant, the islands having few alternative resources to provide a decent standard of living, so people left for a better life elsewhere.

Sometimes out-migration from small islands has not been voluntary and islands affected have not always recovered from resultant major population losses. Thus Gozo, the second island of the small archipelago that is the nation of Malta, has not recovered comparatively since being near depopulated in 1551 when a Turkish raid led by the fearsome Dragut saw some thousands of islanders taken off for ransom or to be sold into slavery. Population estimates for Gozo and Malta before that raid had the smaller island at one third or more of the island of Malta's total (5000 versus 12,000 in 1528; 5000 versus 15,000 in 1530). Two estimates for 1565 show Gozo with 2500 people and Malta with either 19,500 or 17,500 (Biagini, 1974). In the modern era Gozo's population is about one tenth of that of the island of Malta. Nor is this situation likely to change, for Malta has far more in the way of both tourism and industrial infra-structure than what became the backwater island of Gozo. It fact, Gozo has continued to suffer from out-migration, voluntary this time, particularly to the United States and Australia in the decades after World War II. More recently, there has been some return migration, especially of retirees, and many newish houses on Gozo are emblazoned with kangaroos for Australia or eagles for America, to mark the place where their inhabitants made their money. There are benefits to islands from attracting people back, even if only to retire.

Occasionally, natural, economic or socio-political disasters stimulate islanders to leave. There was considerable emigration from Fiji, especially within the Indian community in the late 1980s, after the coup of 1987. A natural disaster was Typhoon Paka, which, followed by an El-Niño-related drought, hit Ail-inglaplap and other outer islands in the Marshall Islands in 1997 and 1998 and saw fresh emigration (*Pacific Islands Monthly*, May 1998). The eruption of the Mount Soufrière volcano on Montserrat in the mid- to late-1990s was another natural disaster that saw many islanders leave.

Out-migration may be only part of the response to changing circumstances. Mauritius provides a useful example of an island where, in the face of economic difficulties, it was the economy that was adjusted as well as the population. It was sugar that caused the French, and then, after 1810, the British, to bring people to the island, which had been unpopulated when first discovered by the Dutch in 1598. By the end of the nineteenth century first slaves and then indentured labourers had helped to bring about a reasonable balance between population and resources. In the post-World War II period population growth began to outstrip the island's potential ability to support its people. However, as Chapter 8 will examine in detail, the authorities reacted by not just setting up a family planning programme which, with out-migration, reduced population growth, but also by diversifying the economy from sugar into tourism, finance and, especially, manufacturing.

The French possession of Saint-Pierre et Miquelon off Newfoundland provides an example where the population remained *in situ*. These islands were France's 'booby prize' after the Seven Years' War in 1763, but the British invaded and deported the people in 1793, the islands being finally returned to French control at the Second Treaty of Paris in 1814. Re-populated, largely by French Basques and Normans after 1816, the islands' location by the Grand Banks ensured that their principal economic activity was fishing, especially for cod. The islands are metaphorically and, in the case of the now abandoned tiny Ile aux Marins off Saint-Pierre town, actually, a museum to cod fishing. There are no more cod to catch – Canada, especially from Newfoundland; France from her mainland as well as Saint-Pierre and the fleets of many other nations have fished out the area. Canada imposed a two-year moratorium on cod fishing in 1992, which has since been extended indefinitely. With the exception of some tourism, the disappearance of the cod, the one resource that drew people to these barren rocks, has left Saint-Pierre in a parlous position economically.

However, there has not been much of an out-migration response because the French government provides extensive support to the colony and its people *in situ*, as a price worth paying to maintain this piece of France in North America.

Not every case of an economic downturn thus leads to significant emigration from a small island; there have on occasion been alternative strategies and outcomes. It is common, however, to see economic/social/environmental pressures being associated with emigration from small islands and the impact of this will now be discussed.

Island out-migration and its effects

Out-migration can be of immediate and direct benefit to island populations. There could be more chance of the stayers getting a job; domestic resources are shared between fewer people, population pressure is relieved. In addition, those who have left may well send remittances to family members left at home on the island. A sum of about Aus$5 m is remitted each year to the benefit of Kiribati from I-Kiribati overseas. People on Cape Verde in the Atlantic are helped by remittances from about 400,000 people from these islands who live abroad, mainly in Portuguese-speaking Africa, Lisbon and Boston in the USA, as many as live at home. 'The goal of every Cape Verdean is first to survive, then to emigrate' (McGirk, 1989, p. 48). More people from Tokalau live in New Zealand than in Tokalau itself. These island diasporas support many of the folks back home and remittances have become a recognised way of sustaining life and society in many islands. Thus the word 'remittances' is built into the term MIRAB (MIgration, Remittances, Aid, Bureaucracy) used to describe some South Pacific economies (see Chapter 8).

However, out-migration, particularly if sustained, particularly if acting upon a small population base, may instead be detrimental rather than beneficial to the sending island's society and economy. There are two related problems. Firstly, most migrants tend to be young, which means that the population left behind will be an ageing one. The second problem is that in the limiting case, emigration may take a small island's population below the threshold needed to sustain essential services such as schools or shops. What is an island's population to do if there are no longer enough people to justify the provision of services, to provide marriage partners or, in an extreme case, to crew a boat or plant the crops?

These problems can be considered in both historic and modern periods. The tiny population and extreme isolation of Tristan da Cunha make the constraints of insularity here as extreme as anywhere in the world. In the nineteenth century the population twice came close to falling below viability. In the 1850s, following the death of William Glass, the island's founding and principal resident, many of his family members moved away from the island. By 1857 only 28 people in four family groups remained and continued occupation was in doubt, though the islanders struggled on, boosted later by some of the Glass family returning. In 1875 when *H.M.S. Diamond*, under command of Captain

G. Stanley Bosanquet, visited Tristan, her chaplain performed a wedding and baptised 25 children. Bosanquet took a census and found the population to be 85. He had been ordered to ask if the people wished to evacuate their island and, although the matter was debated, the islanders chose to remain, only three of the 14 families expressing a desire to leave.

Ten years later, in 1885, 15 of Tristan's 18 able-bodied men were lost when the island's longboat put out to make contact with a passing ship and was never seen again. After this a number of bereaved families left and continued habitation was again in doubt. Then in 1892 two Italians, Andrea Repetto and Gaetano Lavarello were shipwrecked and remained on Tristan rather than face the sea again. More returning migrants, two with new wives from Ireland, arrived in a group of 17 in 1908 and the island's population began its climb towards the present total of about 300, now fairly comfortably supported by controlled exploitation of local crayfish reserves (Royle, 1997).

Out-migration from the Irish islands

By contrast on many small islands depopulation has occurred. A study by Royle and Scott (1996) demonstrated that regarding those off island, the least accessible islands were most likely to lose their population unto abandonment (Table 5.1). In 1841, 211 Irish islands were home to 38,138 people; in 1991 only 66 islands were still populated and, together, housed under 9700 people. One island population that was lost was that of the Copeland Islands off County Down. In the nineteenth century there was a fairly stable economy supporting six farming families on Great Copeland with a maximum population of 46 in 1881. Towards the end of that century greater opportunities and an easier life elsewhere drew many of the people from the island. By 1911 there were only three families left, giving a population total of 25; the school had closed and not one of the adult children was married. Fourteen people between the ages of 18 and 30 lacked partners; perhaps consanguinity had become regarded as a serious issue. Certainly all the families were related; two had the same surname.

Table 5.1 Population of the 72 Irish islands which had more than 100 people at some time between 1841 and 1991

	1841		1991	
	n	*population*	*n*	*population*
Islands with a fixed link	16	6640	19	5961
Islands without a fixed link	56	27,821	27	3570
% recorded population on islands with link	23.9		62.5	

Source: Censuses of Ireland (Royle and Scott, 1996).

Note:
By 1991 only 46 of the 72 islands had a population figure recorded. Some had been abandoned. Others were no longer regarded in the census as 'islands off the coast'.

Under these circumstances depopulation had become inevitable. In 1931 there were two households and six people and the last, lonely inhabitant was recorded at the 1951 census (Royle, 1994).

The most famous example of Irish island depopulation relates to Great Blasket Island off County Kerry on the west coast. Remarkably, from the island's population, which never reached more than 176 (in 1916), emanated three revered Irish-language authors, Thomas O'Crohan, Maurice O'Sullivan and Peig Sayers. All three penned autobiographies of their lives on the island which enables readers to re-create the society there from the late-nineteenth century to the mid-twentieth centuries (O'Crohan, 1929; O'Sullivan, 1933; Sayers, 1936; 1939; see also Mac Conghail 1987; Royle, 1999a). Life was hard. O'Crohan (1856–1937), the earliest of the three, details the way in which islanders earned their living – farming, fishing, trading with the mainland and passing ships, gathering wrack (material washed onto beaches) and cutting peat. Islanders built their own houses, too. In fact the occupational pluralism revealed is very reminiscent of that detailed for the Aran Islands generations earlier in Chapter 4. O'Crohan's book, *The islandman*, has bleak passages, especially those on the deaths of some of his children, but nonetheless he portrays a vibrant culture in a fairly positive way. He was conscious, though, that the way of life he described was passing, 'for the like of us will never be again' (1929, p. 244), as many of his contemporaries were abandoning Great Blasket for a new life in America.

Peig Sayers (1873–1958) had wanted to migrate to America. Instead she married a Blasket man, moved to the island, and made her life 'on this lonely rock in the middle of the great sea – going to bed at night with little food and rising again at the first chirp of the sparrow, then harrowing away at the world' (1936, p. 211). A constant theme throughout her books is emigration, which is seen as an almost inevitable response to the confines of island life in the early and mid-twentieth century when there were so many more opportunities developing elsewhere. Like O'Crohan, she lost a child, Tomás, to a cliff fall on the island, but what caused her almost as much pain was to lose other children such as her sons Pádraig, Micheál and Muiris to America. Muiris was:

> Deeply attached to his country and to his native language and he never had any desire to leave Ireland . . . but he too had to take the road with the others, his heart laden with sorrow. . . . The day he went will remain forever in my memory because beyond all that I had endured nothing ever dealt me as crushing a blow as that day's parting with Muiris.
>
> (1936, p. 185)

Sayers was fond of her island, her 'beautiful little place, sun of my life' (1939, p. 128) but realised that it could not hold the young. As she had said to Muiris, ''Twould be a bad place that wouldn't be better for you than this dreadful rock' (1936, p. 186). Her son, Micheál, returned to Great Blasket, but to do so was to have failed, he returned because 'of the hardship of the world' (1936,

p. 188); a brother of O'Crohan had also failed in America and he, too, had had to come back.

The third author by seniority, Maurice O'Sullivan (1904–50), wrote a book that was lighter in tone, about a young person's experiences on the island. His book's English title is *Twenty years a'growing*, and it details his upbringing. Two generations down from his great uncle, Tomás O'Crohan, he was almost as much of an occupational pluralist as his relative had been, but there was more of a focus in his time on fishing and thus the usual island economic fragility attached to such a concentration became apparent:

> The chief livelihood – that's the fishing – is gone underfoot, and when the fishing is gone underfoot, the Blasket is gone underfoot, for all the boys and girls who have any vigour in them will go over the sea.
>
> (1933, p. 206)

One chapter is devoted to the 'American wake' of O'Sullivan's sister Maura, migrating to Springfield, Missouri, like most Blasket young people. (The 'wake' element was because it was accepted that those who were left behind would be unlikely to set eyes on the migrant again.) Their father was distraught: 'God help the old people, there will be none to bury them' (1933, p. 218) but Maura's view was 'don't you see everybody is going now and you will see me beyond, like the rest of them' (p. 219). O'Sullivan himself migrated, to Dublin to join the guards (police). His book ends with a return visit to Great Blasket to find 'green grass growing on the paths for lack of walking ... the big red patches on the sandhills made by the feet of the boys and girls dancing – there was not a trace of them now' (1933, p. 298).

The Blasket authors revealed that migration was an inevitable response to the pressures of island life – a life that could not meet up to twentieth-century expectations. Here, eventually, as with St Kilda when the last 36 people were evacuated together in 1930, there was the drama of the last boat out. This was in 1953 (Figure 5.1) and one of its passengers must have been Peig Sayers, as she did not die until 1958. However, as has been shown, there had been a steady haemorrhage of people for decades before the last boat. On many other islands, such as Gola, County Donegal, depopulation was just a continuation of this slow drift away of islanders until one winter, no-one stayed (Aalen and Brody, 1969). One man who lived through an island's depopulation was historian Patrick Heraughty, once of Inishmurray, Co. Sligo. He recorded that emigration had always been a way of life but had to some extent been balanced by return migration. Then 'in the 1920s and 1930s emigrants no longer returned. Life was better elsewhere' (1982, p. 71). By the 1940s the remaining islanders were petitioning the council to resettle them on the mainland and eight houses were built on the coast opposite the island. However, in the event, only six were needed for the drifting away of people had continued and there were just 46 people taken off Inishmurray at the official evacuation in 1948, mostly old or very young. This was from an island that, at its peak in 1881, had supported 102 people.

Figure 5.1 The deserted village, Great Blasket, County Kerry, Ireland, 1997

Leaving the Irish islands in the nineteenth and twentieth centuries was not without possible drawbacks to the migrants, including loss of identity and position. Irish playwright, Brian Freel, dealt with this issue in *The gentle island* (1973), set in the make-believe island of 'Inishkeen', County Donegal. As the play opens, the residents have voted to abandon the island. One family, that of Manus Sweeney, refuses to leave. Manus says of the departing islanders passing his cottage on the way to the boat taking them from Inishkeen forever:

> They belong here and they will never belong anywhere else! Never! D'you know where they're going to? I do. I know. To the back rooms in the back streets of London and Manchester and Glasgow. I've lived in them. I know. And that's where they'll die, long before their time – Eammon and Con and Big Anthony and Nora Dan who never had a coat on her back until this day. And cocky Bosco with his mouth organ – this day week if he's lucky he'll be another Irish paddy slaving his guts out in a tunnel all day and crawling home to a bothy at night with his hands two sizes and his head throbbing and his arms and legs trembling all night with exhaustion. That's what they voted for and if that's what you want, it's there for the taking.

Irish island depopulation has not yet finished; 27 islands (excluding 8 lighthouse islands now unmanned) from the 1971 total of 94 inhabited islands had become uninhabited by 1991. These included Rutland Island, County Donegal, which was mentioned above, and Inishbofin, County Donegal. This island the author visited in 1984, where he had peered in through the dusty window of the schoolhouse to see tumbled desks and a calendar on the wall dated 1981, the year the death knell of the island was sounded when its school closed. A decade earlier, in 1974, the author visited Dursey Island, County Cork by its cable car. At that time the island housed about 30 people and had a post office and a school. In 1997 the author revisited, and, on the now battered cable car, fell into conversation with one of the school's last pupils, for it had closed in the late 1970s through want of pupils. The population was then down to six, all elderly, who formed five households. One of the island's three villages had been totally abandoned; another had just one occupied house. The Post Office and shop had gone; the island had no services of any kind. A couple of houses had been bought by foreigners for seasonal occupation but it seems likely that once the last of the six permanent inhabitants dies or leaves, Dursey will become another on the long list of abandoned Irish islands. Paradoxically, it may be that its reasonable access provided by the cable car actually hastened Dursey's decline, for the land on the island still farmed is managed by an islander now living in more comfort on the mainland. He commutes to Dursey by cable car to visit his property. Without such ease of access perhaps he and his family would have stayed (Royle, 1999b).

Other cases of out-migration

The fate of the small Irish islands is not unusual. Throughout the world's island realm the more remote islands, those with fewest resources or suffering other economic problems, have undergone or are undergoing relative and/or absolute population decline as their inhabitants sought or seek better lifestyle opportunities elsewhere. For the Hebrides, emigration, a major factor in life up to the 1960s (Hoisley, 1966) continues. A recent newspaper article brought to public attention the continued problems of out-migration with its headline 'Economic exiles depopulating the Western Isles'. 'There's no work. Petrol is dear. Crofting doesn't pay now the prices of sheep and cattle have gone down so far. People go to the mainland and don't come back,' said one islander. The Registrar General expected the population to fall 14 per cent, from 28,000 to 24,000 between 2000 and 2015 (*Independent*, 6 March 2000).

Even relatively large islands were affected. The case of Ireland has been mentioned. Another is Iceland, which lost 20 per cent of its population in the late nineteenth century to emigration and, at that period, its leaders petitioned US President U.S. Grant to allow the entire population to be resettled in Alaska (Hannibalsson, 1999). Iceland is now one of the wealthiest countries in the world, at the Millennium it was fifth in the world in terms of GDP per capita, supported largely by its fishing industry, so it is a good thing that it was not depopulated.

The opportunities emigrant islanders seek may be on their capital islands, neighbouring mainlands, in those few countries still permitting fresh immigration, or with present or past colonial powers. Thus islanders from the then

British West Indies, also Cyprus and, to a lesser extent, Malta, moved in large numbers to the UK in the post-World War II period, starting with the 492 Jamaicans aboard S.S. *Empire Windrush* in 1948. Islanders from the French DOM-TOMs have the right to move to mainland France and many have done so. The diaspora of the Cape Verde Islanders has already been mentioned. Mainlanders migrate, too, of course. Rural to urban migration is rife within developing world countries. International continental-based migration takes place as with, say, Mexicans and other central Americans moving to the USA – legally and illegally. However, it seems clear that certainly since the ending of large-scale European migration to the New World, as a proportion of the population, more islanders are likely to move than mainlanders, and they are also more likely to move outside their region. The reason is clear – the home island's actual or perceived economic and social opportunities are less than those available in the richer and more varied economic circumstances of the places to which they migrate, be these capital islands or continental cities. Think of *West Side Story*, with its depiction of Puerto Ricans in New York, especially the lyrics to the song *America* (1957; book by Laurents, lyrics by Sondheim).

Island immigration

Islands can also be places of immigration. If one recovers its economic vitality with a revival of a bust industry or the assumption of an alternative, then a population influx might be experienced, to replace those who left in a previous economic downturn. The utilisation of many Mediterranean islands for tourism has seen population growth again in some cases, after problems with agriculture saw them lose people earlier to emigration. New employment and money-making opportunities have led islanders to return to such islands; rising population totals also reflect tourists buying property on the islands, for holiday or retirement relocation, and, finally, labour requirements may see new immigration. Thus, for example, on Mallorca, foreign workers are needed both for menial jobs in tourism directly and, also, as a replacement workforce in agriculture, many of the Mallorquin no longer being willing to work in this sector.

Island immigration then, can, be a positive thing, reflecting absolute opportunities. It also contributes to the islands' domestic economy as demand for goods and services must increase. There could be problems, though, in that in a way similar to gentrification or counter-urbanisation, immigrants might distort the local housing market and push prices beyond the reach of locals. On Mallorca, estate agents' advertisements in German can now be observed.

If migration is within a national island or archipelago, then usually the migrant just ends up in the capital city or capital island. There are a number of Third World island nations which now have the unenviable position of a population decline, absolute or relative, in their outer islands matched by a population increase and overwhelming pressure on the environment and infrastructure of their urbanised capital islands. There might also be social problems. In June 1999, a State of Emergency had to be declared in the Solomon Islands because of clashes between migrants from the island of Malaita and people on the capital island of Guadalcanal who accused the migrants of bringing crime with them and also of squatting on traditional homelands. The fact that Malaitans dominate the government, civil service and business on Guadalcanal was another source of tension. The self-styled Guadalcanal Liberation Army had taken up arms.

Elsewhere, problems are perhaps most marked in those nations where the comparative advantage of the capital island is not marked. In St Vincent and the Grenadines, the principal island has much to offer that the smaller Grenadines do not; in French Polynesia, Tahiti is a large, high island and has considerable comparative advantages over atoll groups such as the Tuamotu Archipelago. However, what if the country consists entirely of atolls, such as Tuvalu, Kiribati or the Marshall Islands?

The case of Kiribati

The outer-islands of Pacific atoll countries have few facilities, a very limited infrastructure, though some do have airstrips, and they normally have a rather traditional economy based on the production of taro and other vegetables, fish, and, as a cash crop, coconuts. Many offer their inhabitants little chance of advancement and it is no wonder that the young people often leave, to try their luck in the cash economies of their capital islands. In these countries the urbanised capital island is just another atoll with all the intrinsic merits and domestic resources of the outer-island atoll from which the migrant has come. Funafuti (Tuvalu), South Tarawa (Kiribati) and Majuro (Marshall Islands) struggle to cope.

An atoll is inevitably a difficult location on which to place an urban area, especially as the landmass is not continuous, but is a series of different sized but inevitably thin reef islands (*motu*). Tarawa Atoll has two basic parts, North Tarawa, which is more exposed and remains in discontinuous *motu*, has the economy and society typical of the I-Kiribati outer-islands. South Tarawa is the urbanised area and here the seven *motu* have been linked by bridges and causeways, most recently with the Japanese-provided Nippon Causeway that connected the fairly broad and well-favoured island of Betio to the rest of South Tarawa. This resultant long, thin land mass, 30 km by, perhaps, an average 150 m has basically one road running the length of it. Along this at intervals of usually less than a minute come mini-buses that provide a cheap, efficient public transport service. There are shops scattered along the road and in the evening, especially, the day's catch can be bought from vendors by the roadside – sharing a crowded Micronesian mini-bus with a large tuna is something of an experience. There are shopping nodes at Birkenbau, at Betio and, especially, at the capital, Bairiki, where there is also an open market, a library, government offices and the national sports stadium. At Betio there is the harbour and some harbour industry. At the other end of the landmass is Bonriki International

Airport, with its services to the neighbouring nations of Tuvalu, Fiji, Nauru and the Marshall Islands.

South Tarawa is an overcrowded island under pressure. At every census since the 1960s it has been found to house both absolutely and relatively more of the I-Kiribati (Table 5.2). There were 6101 people there in 1963, 14.1 per cent of the total; by 1995 that had risen to 28,350, 36.5 per cent of the total, living at a density of 1799 per km^2; one of South Tarawa's constituent islands, Betio, had in 1995 6760 per km^2. South Tarawa accommodates such densities not in high rise towers, but in single-storey huts, closely spaced and housing large families. In the case of South Tarawa, 37.3 per cent of dwellings at the 1995 census were simply the traditional 'grass hut' (Government of the Republic of Kiribati, 1997). 'Grass hut' is in inverted commas because the huts here, unlike traditional dwellings on, say, Fiji, are not roofed with grass but with leaves from the pandanus tree. These huts are well adapted to the local conditions. They have large roofs, with gable ends that fan out to provide a maximum shaded area. The sides are open to allow for air to circulate but can be covered by long mats to provide some privacy if not security. Inside the hut there is a raised platform, on which the people live and sleep, this keeps them off the earth (Figure 5.2). The huts are relatively cheap and easy to build and so, if destroyed by storms, can be fairly readily replaced. Many South Tarawa residents, of course, have more substantial housing and there are a growing number of both publicly provided and privately built and owned dwellings with wooden or concrete block walls, though many are still roofed in the traditional way. The population

Figure 5.2 Traditional housing, South Tarawa, Kiribati, 1998

Table 5.2 Population data for Banaba, South Tarawa and the rest of Kiribati, 1931–95

	1931 n	1931 %	1947 n	1947 %	1963 n	1963 %	1968 n	1968 %	1973 n	1973 %	1978 n	1978 %	1985 n	1985 %	1990 n	1990 %	1995 n	1995 %
Total	29,751		31,513		43,336		47,735		51,926		56,213		63,883		72,335		77,658	
Banaba	2607	8.8	2060	6.5	2706	6.2	2192	4.6	2314	4.5	2201	3.9	46	0.1	284	0.4	339	0.4
South Tarawa	3013	10.1	1671	5.3	6101	14.1	10,616	22.2	14,861	28.6	17,921	31.9	21,393	33.5	25,380	35.1	28,350	36.5
Other Islands	24,131	81.1	27,782	88.2	34,524	79.7	34,927	79.7	34,751	66.9	36,091	64.2	42,444	64.4	46,671	64.5	48,969	63.1

Source: adapted from data in the Kiribati 1995 census report.

Note: Banaba (Ocean Island) was exploited for its phosphate deposits until the early 1980s.

is a young one, there are many schools and, at certain times of the day, uniformed, if often barefoot, children crowd the minibuses.

There are problems with sanitation; according to the 1995 census 1100 households used the beach for toilet purposes – on an atoll one is never more than a few score metres from a beach, which is a relief. There are problems over water supply and over other infrastructure. There is a labour surplus; I-Kiribati men are often trained to go to sea on merchant vessels to get them employment off the island. Despite these pressures, South Tarawa seems to cope; the people if anything are overfed rather than suffering from malnutrition; there is a hospital, the schools and provision for going on to further education at the University of the South Pacific. Above all, it is obvious to the most casual observer that the traditions of the people are maintained even in overcrowded, urban South Tarawa. The radio plays largely local music. Gilbertese, now known as I-Kiribati, is heard everywhere, although English is also universally known and is the language of government. Every few hundred metres are *maneaba*, open sided, thatched or metal-roofed halls, which are the community centres. Above all there is the continuation of local traditions regarding food production. The most telling statistic in the 1995 census is the fact that, despite the high population density, 66.3 per cent of the population of South Tarawa were classified as working within the 'village' or food production sector, and this includes the young people. In fact, 68.2 per cent of males aged between 15 and 24 did so – as well as 67.3 per cent of females, above the mean. Almost everywhere there is agriculture. This is not in the form of fields, but of gardens and taro pits and particoloured pigs in sties made from recycled material containing tyres sliced in half lengthways for water troughs. And there are coconuts. This continuation of tradition is not to say that the I-Kiribati would not wish to own Mercedes and brick houses if the opportunity arose. They do not necessarily remain in pandanus leaf huts and share rickety minibuses by choice, but, in the meantime, the I-Kiribati cope reasonably well on their capital island with all the pressures from immigration (Royle, 1998; 1999c).

The contrasting case of the Marshall Islands

The same continuation of tradition cannot be found in the island group north of South Tarawa, the Marshall Islands. This atoll nation suffers from basically the same demographic and migration pressures as Kiribati. The outer-islands lack opportunity and the chance for a fulfilling life in any but a traditional way, so the young migrate. Some make it off the islands, most end up in Majuro Atoll, the capital and one of only two urbanised islands. The other urbanised island is the rather special case of Ebaye. This is a small landmass of the mighty Kwajelein Atoll. On Ebaye live the civilian workers of the American missile range facility based at the atoll. This activity necessitates a large civilian labour force to service the American residences and the military activities focused on Kwajelein Island itself. The Marshallese are not permitted to live on that island (local powerlessness again). Instead they live in overcrowded and rather insanitary squalor on Ebaye and are ferried over each day to Kwajelein. The urbanisation of Ebaye is thus bound up with the American military presence and must be regarded as a special case. So, we will focus instead on Majuro, a capital island in the same way as South Tarawa, Funafuti, Mahé in the Seychelles or Male in the Maldives. Here the situation is infinitely more depressing than on South Tarawa, although there are many similarities, not least in the land formation and the response to it. On Majuro, too, the various *motu* to the south of the atoll are linked up by a single road (in 1998 it was upgraded courtesy of Japanese aid) which passes over bridges and causeways, some built by the Japanese. Along the road run communal taxis rather than minibuses, but the service is equally frequent, if not quite so cheap. There are uniformed schoolchildren; many schools are needed for the young population, some of whom go on to the University of the South Pacific, too. There are shops and supermarkets, better shops than in Kiribati – Air Marshall Islands offers Christmas shopping trips from South Tarawa to Majuro. There is a considerable harbour here and more industry than on Tarawa; Tabolar, a copra-processing firm, has a large facility and locally-produced copra manufactures such as coconut oil skin care products and soap are available. There is a brewery and a desalinated water bottling plant. There is also multi-channel television run by a cable company, which broadcasts week-old American programmes including news bulletins. There is a recording studio, which produces tapes and CDs of Marshallese music, which makes up a considerable part of the local radio output. As with the I-Kiribati, the Marshallese are proud of their culture and also of their recent independent political status as the Republic of the Marshall Islands (though they remain in 'free association' with the USA, as explained in Chapter 3). Their political status is marked by a large and expensive 1993 government building, of international appearance, which houses the assembly, the *Nitijela*. This is in complete contrast to the *Maneaba ni Maungatabu*; the House of Assembly on Tarawa, which was built by the colonial British in 1964 and is modelled on the architecture of the local community buildings, the *maneaba*.

There are the same population pressures on Majuro as on other urbanised capital islands. Table 5.3 shows that the population of Majuro rose almost nine times from 1958 to 1998 to reach an estimate of over 30,000 whilst its proportion of the Marshallese population doubled to reach almost half. The early German colonialists had been based on Jaluit Atoll. Nor did the Japanese, who ruled the Marshall Islands after the Germans, develop Majuro to any great extent. It was only in World War II that Majuro began to be built up, rather hastily, given the exigencies of wartime, by the Americans who had pushed the Japanese from the island chains. Their need was for an airstrip and a harbour, and part of the atoll was laid down for an airstrip as already described. The harbour facilities were best at the eastern extremity of the lagoon, far from the traditional Marshallese focus on Majuro which was at Laura in the extreme west. Many Marshallese, of course, flocked to the island given the opportunities of the American presence and the urban area of Darrit-Uliga-Delap (D-U-D) began to form. The residences there were never traditional Marshallese in

Table 5.3 Majuro and the Marshall Islands population statistics, 1958–98

	1958		1967		1973		1980		1988		1998 (est)	
	n	%	n	%	n	%	n	%	n	%	n	%
Total	14,163		18,799		25,045		30,873		43,380		62,924	
Majuro	3415	24.1	5249	28.9	10,290	41.1	11,791	38.2	19,664	45.3	30,204	48.0
Other islands	10,748	75.9	13,370	71.1	14,755	58.9	19,082	61.8	23,716	54.7	32,720	52.0

Source: Republic of the Marshall Islands, 1997.

Note: The most recent census data as this table was prepared was from 1988. The 1998 figures are estimates. The CIA estimated the 2000 population to be 68,126 (http://www.odci.gov/cia/publications/factbook/geos/rm.html#People).

design; nor are they now. There are some local features; occasionally houses have shutters rather than windows, for example, but the housing is rather non-descript and non-traditional in appearance (Figure 5.3). Into this crowded (over 11,000 per km² in D-U-D) environment squeeze many residents from the outer-islands as well as those whose families have now lived on Majuro for a couple of generations and whose numbers increase – the birth rate was 46.6 per thousand at the 1988 census. In addition, it is clear from both observation and government reports and statistics that the over-urbanisation of this atoll has brought in train social and welfare problems of a sort not so evident on South Tarawa. Majuro was not the traditional major island of the Marshall group. The difference between Majuro and Tarawa is that the urban migrant Marshallese have largely given up their traditional way of life:

> Young people growing up in the urban centres learn virtually no traditional skills that previously defined a person as a Marshall Islander. Island medicine, fishing techniques, how to climb or tap a coconut tree, how to prepare and how to preserve pandanus and breadfruit are some of the multiplicity of skills that previously provided a person with self-esteem and pride. In the meantime, the western education system is not nurturing life skills adequate for a majority of islanders to succeed in a modern economy.
> (Government of the Marshall Islands, 1996, p. 7)

The lack of food production in D-U-D is immediately obvious and is in striking contrast to South Tarawa. There is still traditional agriculture including taro pits

Figure 5.3 Housing, Darrit-Uliga-Delap, Majuro, Marshall Islands, 1998

as well as coconut exploitation at Laura, but in D-U-D, apart from a few pigs rooting about, one sees little sign of production. Statistics of coconut production confirm the story. Production in the Marshall Islands as a whole changed from an index of 100 in 1974 through 118 in 1986 to 110 in 1996. That for a typical outer-island such as Wotje went through 120 to 181 for the same years, but for Majuro production fell away, through 54 in 1986 to 47 in 1996 (Republic of the Marshall Islands, 1997).

One result of the lack of local food production on Majuro means that there is little local produce available to buy, fresh foodstuffs tend to be imported and are both expensive and sometimes of poor quality, given the remoteness of these islands. The local fish can be expensive and sometimes in short supply. The result is that a lot of Marshallese on Majuro rely on inexpensive processed food – tinned corned beef is particularly popular. Their diet is high in fat and short on some of the essential nutrients and vitamins, and there is considerable obesity amongst the Marshallese on Majuro. In addition there is a very high rate of diet-related diabetes and a large proportion of hospital admissions are related to this disease. Further social and welfare problems relate to the breakdown of the old social systems, in particular the extended family system.

> Traditional leaders point out that younger people nowadays don't know their genealogies and don't realise the importance of their relatives. Many youths don't know the meaning of their *jowis* [extended grouping of clans] and knowing their own *jowi* and the *jowis* of their grandparents is vital to knowing who you are related to and what your social relationship is with different people in the society.
>
> (Government of the Marshall Islands, 1996, p. 7)

In short, excessive immigration to and urbanisation of Majuro has led to the population modernising and losing its traditions, but without being able to replace the fulfilment and achievements of traditional life with anything particularly satisfactory in return. Perhaps the reason is the Marshalls' links with America and exposure to this superficially attractive and materially wealthy society, if only through week-old television and their use of the almighty US dollar (in Kiribati, the currency is the less almighty Australian dollar). Many of the Marshallese would wish to live like rich Americans but have little realistic chance of being able to achieve this goal, given the inevitable economic constraints on life in such a resource-poor and remote location. Self-esteem perhaps suffers as a result, for these people who are no longer traditional Marshallese but cannot be proper Americans.

It is ironic, perhaps, that in 1998 there was opened a Marshallese cultural centre, designed to cherish such culture as heritage. The location of this centre was Kwajelein Island, the American base island on which Marshallese can no longer live. In Kiribati, the traditional culture is a living heritage. In many ways, Kiribati is a less modern society than the Marshall Islands. However, as far as an outside observer can judge, I-Kiribati society seems to be more fulfilling in that even in the crowded ecumene of its capital island the local people seem better able to deal with their islandness (Royle, 1998; 1999c; 2000).

Island ethnicity

This chapter has dealt with islands both gaining and losing populations. One result of population movements in the insular situation mentioned but not yet fully discussed is, in some cases, the assumption of an ethnically mixed population. Colonisation is usually the cause. In some cases, for example Fiji, it has resulted in a population containing the aboriginal inhabitants and people who are descended from those brought in by the colonial power. In other cases such as Mauritius and Réunion, there were no aboriginal peoples but the colonisers' population imports were from different racial groups. Elsewhere aboriginal islanders were almost entirely or even completely exterminated either by disease or malice so, again, ethnic mixes were all imported – many West Indian islands fall into this category.

Some islands manage their ethnic mixtures well – Mauritius would be an example here. However, most islands with ethnic differences like, sadly, many mainland areas do not manage well – Bosnia, Kosovo and Rwanda were some of the continental areas undergoing ethnic strife during the period this book was written. On islands, perhaps because of the restrictions on space, ethnic rivalry can be particularly troublesome. Sometimes the controversies are petty – disputes on the racial composition of the West Indies cricket team is one such example which, on occasion, crops up. In other times and places the rivalries can be deadly earnest. In Ambon, capital of Malaku Province (formerly called the Spice Islands), Indonesia, there was considerable ethnic strife between Christians and Muslims during unrest that followed the ending of the rule of President Suharto in 1998. In Fiji the rivalry is between native Fijians and Indian Fijians, whose ancestors were brought in as indentured labour to work in the sugar industry from 1879. Indian Fijians make up 44 per cent of the population, but find it hard to reach the full political power such a position might give them because of the native Fijians' attitude to land ownership and customs which they feel are threatened by the Indian Fijians. In 1987 there was an ethnic coup against an Indian-dominated parliament and a native Fijian, Lieutenant-Colonel Sitiveni Rabuka, seized control. Fiji was expelled from the Commonwealth and became a republic and many Indian Fijians who dominated commercial and professional life left the island, concerned about militant Fijian nationalism. Civilian rule was restored in 1990, although Rabuka remained as a government minister. Fiji (now officially called the Fiji Islands) was readmitted to the Commonwealth in 1997. However, in May 2000 there was another coup staged by native Fijians against the government then led by an Indian Fijian. The Prime Minister, Mahendra Chaudry, who had been democratically elected in 1999, beating Sitiveni Rabuka, was deposed during a 55-day period of captivity when he and 17 other MPs were amongst 31 hostages held in the parliament building by ethnic Fijians led by George Speight. Speight failed in his ambition to become Prime Minister himself. Fiji was suspended from the Commonwealth once more. More seriously, perhaps, its sugar-based economy was 'crippled' by sanctions imposed by major trading partners, Australia and New Zealand, and 'the country's other major earner, tourism, has been devastated' (Crisp and Bohane, p. 28).

Small island populations: the future

Many small islands face continued difficulties from out-migration with their continued habitation being in doubt. 'Pitcairn, the end ever nigh' and *'Pitcairn cherche jeunes hommes desperement . . .'* (Pitcairn searches for young people desperately . . .) are two article titles that convey this message (Connell, 1988, and Anon, 1998, respectively). The problems of providing a satisfactory economy and social life from insular environments remain. Haitians, for example, continue to try to leave and, despite the dangers, many still take to the sea in attempts to reach America as illegal migrants. Around 40 Haitians were drowned in March 1999 when their boats capsized off Florida.

However, as stated, there is now much outside support for islands and that can help. Scottish Natural Heritage, the conservation body which owns the nature reserve island of Rum in the Scottish Hebrides, is looking to bring the population up from its present 30. The aim is not to return it to anywhere near the 400 who lived there before the Highland Clearances saw most of the population shipped off to Nova Scotia (quite probably to Cape Breton Island). Instead 'we want to make sure there are children in the school and that the population gradually rises to 50, without compromising the reserve'. Scottish Natural Heritage reported that 'lots of people' were interested in moving to Rum, many of them from the South of England (*Independent*, 23 November 1999). This exemplifies a modern trend with regard to islands. Islands are romantic; Rum especially, with its wonderful scenery, wildlife and great house, Kinloch Castle, erected on George Bullough's 'Rhum' (see Chapter 1). People, especially from the overcrowded South of England, can look longingly at a place like Rum and want to move there. They may not be realistic and, if they have to make a living from the island, may find their romantic dreams dashed. However, there are now many people in the western world who no longer have to make a living. The baby boomer generation is moving towards retirement. Many people consider spending their retirement or at least their holidays on the romantic islands and so buy property. Tourist islands, especially, are prone to the development of retirement communities. Mallorca is a prime example with some settlements in the island now given over to retired Germans, Swedes or Britons, who spend at least part of the year there.

On smaller islands, those threatened by population decline, some can assume a new population structure, which can counteract any tendencies for native islanders to leave. Thus on the island of Egilsay in Orkney, a newspaper headline some years ago read 'Southerners go to work on Egilsay' and the article noted that only one household of the 13 contained an original islander. The

Figure 5.4 Welcome to Skopun, Sandoy, Faroe Islands, 1999

new islanders were from mainland Scotland, Ireland, England and Germany. One was a Rastafarian. One family, from Nottingham, was learning to look after 117 sheep and a goat – from a book. They moved to Egilsay to be able to bring up their children in safety (*Independent*, 8 July 1989). Even Dursey Island, inevitably now to be abandoned by its traditional population, has holiday homes and will house people from continental Europe at least for a few weeks per year. On one recent summer visit to Sherkin Island, County Cork, Ireland, the author observed a helicopter parked on the front lawn of a house, a strong indication that the resident was not from the traditional farming/fishing community of the island. Such immigration may keep some small islands inhabited, but the traditional relationship between domestic resources, population and habitation will have broken down. And the locals may continue to take a different attitude to their own islands than the romantic incomers. On Sandoy Island in the Faroes is the village of Skopun. On its road sign appears a hand-written addition, in only slightly misspelled English: 'Wellcome (sic) to Hell' (Figure 5.4).

References

Key readings

The edited volume by King and Connell presents a variety of studies on small island migration from around the world. The book by Aalen and Brody studies one island through its last days. The pressures of migration on small island society in the twentieth century is illustrated by the three Great Blasket autobiographers; it is especially prominent in Sayers's books. The contrasting problems of excessive immigration are well revealed by the UNICEF-published study on the Marshall Islands.

Aalen, F.H.A. and Brody, H. (1969) *The life and last days of an island community*, Mercier Press: Cork.

Anon (1998) 'Pitcairn cherche jeunes hommes desperement . . .', *TAHITI-Pacifique*, 84, pp. 26–7.

Biagini, E. (1974) *Le Isole Maltesi*, Accademia Ligure di Scienze e Lettere: Genova.

Connell, J. (1988) 'The end ever nigh; contemporary population change on Pitcairn Island', *Geo Journal*, 16, pp. 193–200.

Crisp, P. and Behone, B. (2000) 'The storm before the storm? The coup may be over but little dust has settled', *Asiaweek*, 28 July, p. 28.

Freel, B. (1993) *The gentle island*, Gallery Books: Oldcastle, County Meath (first published 1973).

Government of the Marshall Islands (1996) *A situation analysis of children and women in the Marshall Islands*, UNICEF Pacific: Suva.

Government of the Republic of Kiribati (1997) *Report on the 1995 Census of Population*, Statistics Office, Ministry of Finance: Tarawa.

Hannibalsson, J.B. (1999) 'The raison d'être of the Icelanders', *Iceland Review*, February 1999, pp. 28–33.

Heraughty, P. (1982) *Inishmurray: ancient monastic island*, O'Brien Press: Dublin.

Hoisley, H. (1966) The deserted Hebrides, *Scottish Studies*, 10, pp. 44–68.

Jones, H. (1993) 'The small island factor in modern fertility decline: findings from Mauritius', in D.G. Lockhart, D. Drakakis-Smith and J. Schembri (eds) *The development process in small island states*, Routledge: London, pp. 161–78.

King, R. and Connell, J. (eds) (1999) *Small worlds, global lives: islands and migration*, Pinter: London.

Kennedy, L., Ell, P.S., Crawford, E.M. and Clarkson, L.A. (1999) *Mapping the great Irish famine: a survey of the famine decades*, Four Courts Press: Dublin.

Lazaridis, G., Poyago-Theotoky, J. and King, R. (1999) 'Islands as havens for retirement migration: finding a place in sunny Corfu', in R. King and J. Connell (eds) *Small worlds, global lives: islands and migration*, Pinter: London, pp. 297–320.

Mac Conghail, M.D. (1987) *The Blaskets: people and literature*, Country House: Dublin.

McGirk, T. (1989) 'Where the consul is God', *Independent Magazine*, 16 December, pp. 48–55.

Mokyr, J. (1983) *Why Ireland starved: a quantitative and analytical history of the Irish economy, 1800–1850*, George Allen and Unwin: London.

O'Crohan, T. (1978) *The islandman*, Oxford University Press: Oxford. (First published as Ó Criomhthain, T. (1929) *An tOileánach*, Oifig an tSoláthair: Baile Átha Cliath.)

O'Sullivan, M. (1953) *Twenty years a'growing*, Oxford University Press: London. (First published as Ó Súilleabháin, M. (1933) *Fiche Blian ag Fás*, Clólucht an Talbóidigh: Baile Átha Cliath.)

Republic of the Marshall Islands (1997) *Marshall Islands statistical abstract 1996*, Office of Planning and Statistics: Majuro.

Royle, S.A. (1994) 'Island life off County Down: the Copeland Islands', *Ulster Journal of Archaeology*, 57, pp. 172–6.

Royle, S.A. (1997) 'The inside world: Tristan da Cunha in 1875', *Journal of Historical Geography*, 23, 1, pp. 16–28.

Royle, S.A. (1998) 'Health and welfare in face of insular urbanisation: the Pacific atoll nations of the Marshall Islands and Kiribati', in P. Santana (ed.) *Health and health care in transition*, International Geographical Union Commission on Health, Environment and Development: Coimbra, Portugal, pp. 113–22.

Royle, S.A. (1999a) 'From the periphery of the periphery: historical, cultural and literary perspectives on emigration from the minor islands of Ireland' in R. King and J. Connell (eds) *Small worlds, global lives: islands and migration*, Pinter: London, pp. 27–54.

Royle, S.A. (1999b) 'From Dursey to Darrit-Uliga-Delap: an insular odyssey. Presidential address to the Geographical Society of Ireland', *Irish Geography*, 32, 1, pp. 1–8.

Royle, S.A. (1999c) 'Conservation and heritage in the face of insular urbanisation: the Marshall Islands and Kiribati', *Built Environment*, 25, 3, pp. 211–21.

Royle, S.A. (2000) 'Population and resources in contrasting environments part 2: The Republic of the Marshall Islands', *Geography Review*, 13, 5, pp. 30–3.

Royle, S.A. and Scott, D. (1996) 'Accessibility and the Irish islands', *Geography*, 81, 2, pp. 111–19.

Sayers, P. (1974) *Peig: the autobiography of Peig Sayers of the Great Blasket Island*, Talbot Press: Dublin. (First published as Sayers, P. (1936) *Peig*, Clólucht an Talbóidig: Baile Átha Cliath.)

Sayers, P. (1978) *An old woman's reflections*, Oxford University Press: Oxford. (First published as Sayers, P. (1939) *Machtnamh seana-mhná*, Oifig an tSoláthair: Baile Átha Cliath.)

6 Islands
Communications and services

Human needs are basically the same the world over. Food must be available to sustain life; shelter from the elements must be provided and there must be a series of social support measures including health and education in order that each generation can bear and rear its children. It is the way such things are provided that varies, according to different opportunities and traditions around the world. This chapter focuses on how these social support measures can be provided in the island situation, because insularity imposes its own difficulties here as in every other aspect of life. Perhaps the most important service activity in an island situation is communications, consideration of which forms the first part of the chapter. Other sections deal with retailing, administrative services and the provision of education and health systems.

Communications

Transportation

Transportation facilities are vital to island living. Telecommunications, which ease isolation, are part of the communications picture and will be considered below, but first we should discuss transportation, the mode of physically moving people and their goods to and from the island. John Donne famously said that 'no man is an Iland, intire of itself, every man is a peece of the Continent, a part of the maine' (1624, pp. 415–16); to extend his reasoning, no island is an island either. By this is meant that there is no inhabited place, no matter how remote that has not had and does not have need for connections to what the ultimate islanders on Tristan da Cunha call the 'outside world'. In the past, islands may have been cut off for long periods. After its colonisation by the mutineers from the *Bounty* and their Tahitian companions in 1790, Pitcairn was not visited for 18 years. The Polynesian inhabitants of Rapa Nui (Easter Island) were cut off for centuries. Their timber resources gone, islanders had no means of voyaging from their home, which they considered to be the navel of an otherwise wet world. It may be that memories of there being an outside world had faded from the people before outsiders came to them again in 1722. These days are gone; Easter Island is an emergency landing strip for the NASA space shuttle, hardly an indication of its being cut off from the world. Rapa Nui was even in the past very unusual, for Pacific Islanders were inveterate travellers, making long and dangerous voyages between the tiny specks of land in that seemingly boundless ocean. Islands are normally poorly resourced and islanders almost invariably needed to trade to obtain not just luxuries, but any necessities that could not be produced from limited domestic resources. On the Marshall Islands the navigators relied on charts made from sticks which taught their people the way to navigate by currents (in addition to by the stars) between their tiny, very low-lying atolls.

In the present day, transport and communications are simply the most important components of island life. On islands people consider and talk about their ferries or air connections in a way and with a continued interest that else-where might be reserved for the weather. This was summed up well by a Gola islander interviewed in Aalen and Brody's classic book on the depopulation of that Irish island: 'everything is handy except you have to be going in and out. That's the only bloody trouble' (1969, p. 126). Island transport systems come up against the familiar problems of scale and isolation. There is a cost, maybe a large one, to overcome the water barrier around the island. But the cost must be met; an island in the modern world without a reliable transport system faces severe difficulties.

An analysis of population history on the Irish islands discovered a clear rela-tionship between population decline and levels of accessibility, and also that every island wanted its accessibility to be improved (Royle and Scott, 1996). In recent years considerable effort has been put into better ferry systems for the Irish islands, including Rathlin in Northern Ireland which has been served by the Scottish ferry company, Caledonian MacBrayne, since 1997, providing for the first time a proper roll-on, roll-off ferry service for this island. Only islanders are normally allowed to take vehicles on board, however, a policy that seeks to preserve the tranquillity of Rathlin as well as taking cognizance of the state of the island's roads. Caledonian MacBrayne (Figure 6.1) is an interesting study of island transport. This company provides almost all the ferry services to and within the islands of the west coast of Scotland; it operates to 23 Scottish islands as well as running several Scottish mainland ferry links and, as men-tioned, that to Rathlin Island. Only a few of the small off-islands of the Western Isles, such as Jura and Luing which have local authority ferries, and some small islands, such as Kerrera and Eriskay which rely on private ferries, are outside the Caledonian MacBrayne empire. The author was taught a poem on Lewis, which is apposite here (he has no idea if it has been published):

> The earth is the Lord's
> And all it contains,
> Save the Western Isles –
> They are MacBrayne's.

The logic of one company dealing with such a plethora of ferry routes is clear to see; this is a prime example of insular scaling-up. The company benefits from

Figure 6.1 The Caledonian MacBrayne ferry to the Isle of Bute, Scotland, UK, 1994

economies of scale in everything from the purchase of fuel oil to crew training. It serves 52 ports and transports about five million passengers and one million cars each year. In fact, the cumbersome name of the company, Caledonian MacBrayne, is a legacy of its history as two, once separate, private companies, the Caledonian Steam Packet Company and MacBrayne's Shipping, neither of which by the 1970s was able to operate profitably. So in 1973 for economic reasons and to secure shipping services for the islands as a social necessity, the two companies were merged and nationalised. There is no competition on almost all its routes, except where air services are available, but even so, despite its near monopoly position in the performance of a vital service, the company has had to be kept in public ownership. The services to Orkney and Shetland from Aberdeen are privately owned, and the privatisation of Caledonian MacBrayne was certainly considered by the British Conservative governments of 1979–97. Their policy was to sell off as many state assets as possible. This raised tremendous opposition on the islands potentially affected; transportation provision is very contentious politically in the Scottish island realm (Didier-Hache, 1987). Delegates from all 23 islands served by Caledonian MacBrayne met at a congress in Oban in 1988 to oppose 'this political ideology, which is being taken to its ludicrous extreme to the great detriment of the way of life of island communities' (Angus Graham, Western Isles Council, *Oban Times*, 10 November 1988). The islanders feared a reduction in government subsidy for services, which would see fewer sailings. Also, they were concerned about the possible break up of the company, perhaps to the detriment of the smaller islands which benefit most, relatively speaking, from the economies of scale possible in the operation of the then integrated system. There was 'overwhelming support for a motion calling for an immediate halt to any moves towards privatisation and the fullest possible consultation with islanders' (*Oban Times*, 10 November 1988). In face of such opposition, the Government commissioned a report from consultants whose conclusion, it was reported, was that privatisation would lead to few savings, if any (*Independent*, 28 September 1994). In 1994 the Government announced that it would retain Caledonian MacBrayne as a subsidised state monopoly. The realities of island life had overcame the beliefs of what, by that time, was a fairly dogmatic government.

In fact, island transportation is often in the hands of governments – the one ship which services St Helena is government owned; internal air transport to the farm settlements of the Falkland Islands is through the Falkland Island Government Air Service (FIGAS). FIGAS is another good example of scaling up. Passengers state to which airstrip they wish to go; FIGAS contract to take them there but, by no means necessarily directly, for a routing is calculated for each day's work which makes optimum use of the aircraft and minimises the distance it has to fly. The author has flown FIGAS, and a return journey between Stanley and Pebble Island saw him land at Mount Pleasant air base, Hill Cove, Chartres and Saunders Island. He had to wait on Saunders Island whilst the plane took a passenger to Westpoint Island, the airstrip there being regarded as too risky to take people in who are not disembarking. Given the pastures and beaches on which landings were made elsewhere, this must be a difficult landing indeed! FIGAS is aware that such round robin journeys are inconvenient to passengers but is finding the provision of a scheduled service 'a difficult nut to crack . . . a realistic and practical approach to this issue will prevail' (Falkland Islands Government, 1998). Governments have often to subsidise island transport; the amount of traffic means that fully commercial operations are not feasible. In 1996–7 aviation revenue to the Falkland Islands Government was £890,579; expenditure on aviation by the government was £1,765,517; estimates for 2000–1 were £1,010,250 revenue and £1,847,820 expenditure (Falkland Islands Government, 1998).

Despite the improvements in recent years in technologies and delivery systems, and recognising the importance of transport to small islands, such transport often retains an old fashioned informality, not to say unreliability. Any person regularly travelling in an island setting will have stories of delays, cancellations and utter frustration. Thus islanders throughout the world press for better and improved transport systems. Inishbiggle islanders in County Mayo, Ireland, who are connected to their neighbouring large island of Achill only by a *curragh* (a traditional small open boat) which carries the post, want a cable car. Residents of Dursey Island, which has a cable car, want a bridge; people on Achill Island who have a bridge would like a railway. Tory and Clare Islands, both of which have recently had their ferry services transformed, would like airstrips; the Aran Islands, which have airstrips and very regular ferry services, want more flights (Royle and Scott, 1996). On Fogo Island off Newfoundland there is a campaign, featuring cartoons, some on T-shirts, to replace the ferry service by a causeway, presumably making use of some islets between Fogo and the mainland. The cartoons feature one 'John Cawsway', in one image shown hugging none other than Saddam Hussein of Iraq. Saddam had been through the Gulf War and lived to tell the tale; Joe had survived the Fogo Island ferry (Figure 6.2).

In the 1980s and 1990s considerable sums were expended to connecting a number of islands through fixed links. These could certainly cause controversy through the high costs of construction, concerns about the impact of the links on the environment and also the way in which fixed links take away an island's special qualities and render it a cul-de-sac. The Seikan Tunnel linking Hokkaido and Honshu in Japan, for example, was originally estimated to take ten years and cost £60 m; instead it took 24 years and just under £3 billion by the time it opened in 1988. The decision to build the Confederation Bridge to Prince Edward Island, a mighty 13.2 km, £460 m structure, which opened in 1997, was by no means universally popular (Begley, 1993). Some people on Prince Edward Island still refuse to use it in protest. Financial problems also relate to

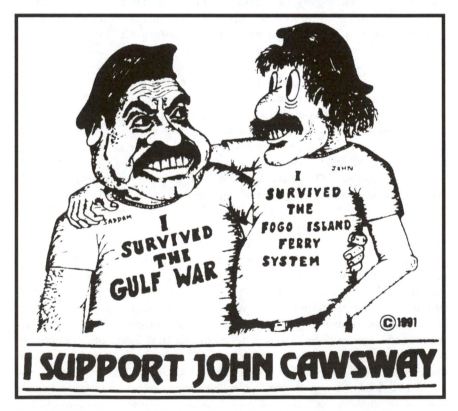

Figure 6.2 John Cawsway and Saddam Hussein, Fogo Island, Newfoundland, Canada
Source: Gerald Freake, on T-shirts sold on Fogo Island, 1995

the cost of using these links. The toll to cross the 1985 Penang Bridge linking Penang to mainland Malaysia costs the average worker about a day's pay. The 1995 Skye Bridge in Scotland has also been controversial because of the high cost of its toll. On Prince Edward Island a seasonal ferry still runs from the other end of the island to the bridge and there are several flights a day to and from the island. By contrast, Skye lost its year-round ferry link to the mainland (though it still has a seasonal service) and so to protest against the bridge by not using it is difficult. Instead protesters refuse to pay the toll and some end up in court, ironically having to use the bridge to travel to Dingwall for their hearings (e.g. *Independent*, 10 February 1996). Despite the controversies, fixed links to small islands are often important economically; Prince Edward Island's tourist trade increased by about one-third during the Confederation Bridge's first year (see also Hamley, 1998). Similarly tourism and retailing increased markedly in Bahrain when the 26 km King Fahd Causeway opened in 1986. In particular, hotels on the island began to fill up at the Muslim weekends. In Denmark, the fixed links from the Jutland Peninsula through Funen to Zealand and from Copenhagen on the eastern edge of Zealand across to Malmo in Sweden, when finished in 2000 created the Øresund urban region. This area of about three million people links Lund, Copenhagen and Malmo, and forms the eighth largest urban region in Europe.

Transportation can make or break an island. St Helena is dreadfully handicapped economically and socially by not having an airport and being one of the few places left in the world where travel times to anywhere is measured in days. Hence the vigorous campaign by islanders and others to have an air service provided, a campaign that may yet bear fruit. By contrast, Batam Island, Indonesia is just off Singapore and is well connected to that island by ferry services of under an hour and by air. This is one of the reasons why Batam was subject to a remarkable industrialisation in the mid-1990s. This story will be taken up in Chapter 8.

Telecommunications

Island life is subject to the tyranny of distance. Offshore coastal islands are peripheral by definition; oceanic islands are even more isolated. The problems of peripherality are compounded by the cost and inconvenience of the journey across the water to reach or leave the island. In such circumstances the advent of telecommunications technology is a godsend. Here is a system of communications which, once set up, is not affected in any practical way by isolation. Costs of operation may reflect distance insofar as a long distance phone call is more expensive than a local call, but the offshore island is probably no worse off than its nearby mainland in this regard. To ring Dublin from the Aran Islands costs no more than from Galway City, whereas to visit Dublin or to obtain goods and services other than by post from Dublin inevitably costs more from the islands. For more isolated islands, however, any call off the island is long distance. Recent advances in telecommunications, including e-mail and the Internet, may be less affected by distance charges.

Telecommunications have become essential to modern life. As recently as 1983 a newspaper article reported that the Pitcairn islanders depended on contact with the outside world through the good offices of a short-wave radio ham in Surrey in England who made regular contact with an islander (*Sunday Times*, 24 July 1983). Such ad hoc arrangements are not now necessary and most islands, at least those in or associated with the developed world, as well as many others, have modern telecommunications systems. These systems represent the one way in which islands can compete on an almost level playing field with less remote areas. More and more business is carried out by means of telecommunication; telecommuting, working from home by computer link-ups, is becoming increasingly common. Retailing can be done on the Internet and certain service activities including some forms of education can be accessed on line, too.

Innovative types of employment, such as call centres, which do not rely on physical contact with customers, have also become popular. Island authorities have been jostling to get a share of these markets. A factor that might help here is that the usual economic problems that face islands tend to keep wage levels depressed. Therefore, as long as the island labour force has the necessary skills, if telecommunications-linked jobs can be brought in, such jobs, uniquely freed from the problems of isolation, can be performed cheaply. On a fairly large insular scale both parts of the island of Ireland benefit tremendously from this new type of work. Until into the 1970s, the Irish telephone system was fairly primitive, but now the island bristles with telecommunications equipment. Ireland has become a key partner in much American business in that some back office work for, for example, New York insurance companies, is carried out remotely in the west of Ireland. At the small island scale, almost every inhabited Irish island now has a telecommunications tower. This is not to bring in major call centres – at this level the problems of insular scale kick in, with regard to limited labour availability. However, at least the advantages of modern communications ease island life and business as well as providing the possibility of enabling some people to work remotely whilst based on the island.

Some islands have had a notable role in telecommunications history because of their strategic location. Cape Breton Island in Nova Scotia was the place from which Guiliermo Marconi first sent messages across the Atlantic to Cornwall in Great Britain, and Marconi's facility on Cape Breton is now a National Historic Park. Marconi's associates also worked on Rathlin Island, County Antrim. Some oceanic islands, such as Ascension in the Atlantic, played important roles as landfalls for transoceanic cables. The first cable arrived on Ascension out of Southern Africa in 1899 and the island ended up as a node for a web of cables running north–south and east–west across the Atlantic, operated by Eastern Telegraph Company, later Cable and Wireless. Ascension later became a relay station for radio broadcasting and at the time of writing the BBC still makes use of the island to transmit World Service programmes. It is fitting, therefore, that it is modern telecommunications that represent one of the best hopes for easing island lives and bringing better economic prospects to them.

On the islands of Canada's Maritime Provinces, investment in telecommunications is seen as an important way in which troublesome economies can be diversified. Prince Edward Island (PEI) has largely been associated with primary activities, especially agriculture – particularly for potatoes – and fishing, principally for lobster. Tourism is also important, but is very seasonal here with the island's long winters keeping the summer season to, at most, 12 weeks. The island's authorities are trying to maximise returns and employment from the existing sectors – 'cultivating island solutions' as one government report had it (Round Table on Resource Land Use and Stewardship, 1997) but diversification is called for, too. Diversification into the quaternary sector is one way forward. Thus, in 1988, the island's small university was very pleased to be chosen as the site for the Atlantic Veterinary College, which brought high paid jobs, skills and business to the island, whilst linking in to PEI's pre-existing animal husbandry industry. Another key sector is telecommunications. This is potentially very significant in this island's extensive rural areas – more than half of its people are rural dwellers and the island has only two towns, Summerside and Charlottetown, the capital. Summerside used to have a significant employer in the form of a Canadian military airbase. This closed in the 1980s and facilities left behind have been successfully re-used to develop a growing avionics industry. Much investment in computing, and ICT (information and communications technology) generally, were associated with this transformation, with the educational needs being met by the university and Holland College, the island's technical educator. There are now 28 computing centres across the island, in reach of all the population of an island that has virtually no public transport provision. The provincial government in 1997–8 ran a scheme that made grants available to households to help with the purchase of computers (with Internet access) for the home. In January 1998, PEI became the first Canadian Province to provide computers with Internet access in each of the island's schools. Perhaps it is significant that this investment seemed to have taken precedence over matters such as upgrading and maintenance of buildings and other capital items: a cartoon in the local press contrasted the computer provision to the poorer quality of some other facilities (Figure 6.3).

ICT, the knowledge revolution, is seen as the way forward for this island, small in scale, isolated and remote from the Canadian heartland. One of its Rural Development Corporations, which has taken the new opportunities very much to heart, is that which deals with the island's small Francophone population. This group, largely descended from Acadians who fled the British decision to deport this group from its long established base in Nova Scotia in 1745, numbers about 4000 people and is concentrated in the southwest of the island around the village of Wellington. This is called the Evangeline Region, named for Longfellow's poem of that name about the Acadian expulsion. The Rural Development Corporation, *La Societé de Devéloppement de la Baie Acadienne*, shares its offices in Wellington with the *College de L'Acadie*. The latter aims to provide the Acadians here with the highest quality links into the Francophone telecommunications world, with the expectation that jobs and business

Figure 6.3 Cartoon marking the arrival of Internet access to every classroom in Prince Edward Island, Canada, 1998

Source: Sandy Carruthers, Guardian News, *Prince Edward Island Guardian*, 24 January 1998

opportunities will arise, thereby. One possibility is call centre work, with the Evangeline area able to offer workers who are fluent in both of Canada's official languages. There is competition here, however, from other islands. For example Ile Madame, an island off Cape Breton Island in Nova Scotia, houses another of the scatter of Acadian clusters and already employs people in bilingual call centre work. Even more significantly across the Northumberland Strait from PEI lies New Brunswick, Canada's only officially bilingual province, which with its concentration of Acadians is also investing heavily in ICT facilities. *La Societé de Devéloppement de la Baie Acadienne* is also investing in more traditional projects such as tourism facilities. However, in rural PEI, as in islands elsewhere, ICT and telecommunications work is seen as one important way to diversify. One problem to be overcome in PEI, however, relates to the island's legacy from its traditional fishing and farming economy. This saw young people able to take up employment as soon as they had reached near adult strength and many left school early without developing decent literacy skills. In the rural parts of the island illiteracy runs at about twice the Canadian average. Most young people stay on at school now, but there are still literacy problems associated with those children coming from illiterate households. Illiteracy and a future in the knowledge industry and ICT do not go hand in hand, and there are now major programmes, partly delivered through the Rural Development Corporations, to improve literacy skills (Royle, 2000a).

In the European island scene, too, telecommunications are seen as the way forward. In *Insula*, the journal of the International Scientific Council for Island

Development, based in Paris and supported by UNESCO, was an article on the topic about Hiiuma, an Estonian island. Hiiuma has three Internet centres supported by the United Nations Development Programme. The first centre of this sort appeared in the National Library in the Estonian capital Tallinn, the next three on Hiiuma:

> The island of Hiiuma did not get these 'internet-houses' by chance. For about three years, from the beginning of [the] telecommunication boom in Estonia, the island has been one of the best developing countryside regions in this field. . . . The economies of the island go down, agriculture has nearly collapsed, fishing is facing hard times; and nature tourism, which was expected to start booming, did not start yet. In addition, [there is a] high unemployment rate and young people leaving for the city to find jobs. And, not the least, frequent cut-offs in the bottle-necking ferry lines. Probably the last mentioned is the main reason why the enthusiastic people on Hiiuma look for better communication. Maybe to compensate the insular isolation from the development centres . . . islanders lose contacts. To counterbalance this, the advantage has been taken from the new informational technologies.
>
> (Kokovkin, 1998, pp. 17–18)

Kokovkin went on to identify the practical advantages of the new facilities, which include increased tourism from people attracted to Hiiuma from information they have found on the Internet; and some in-migration to the island 'from intellectuals and professionals' who can work from distance. These include the Estonian novelist, Tònu Ònnepalu, who can research for his work on the Internet whilst living in the wildest part of Hiiuma. A Swedish businessman resides on the island whilst keeping in touch with his contacts across the Baltic by telecommunications. Telecommunications are the future for this island, for its children, it is already here. Kokovkin concludes, after recalling his own childhood where the television was a rarity, how:

> The village children who go to the . . . internet-room to get immediate connection to the net have no surprise. They operate with the computer as if it has always been there. . . . And some other nets, the fishing ones, are drying in the wind.
>
> (Kokovkin, 1998, p. 18)

Singapore is further down the road to complete electronic coverage than Hiiuma. Its *Vision of an Intelligent Island* strategy of 1992 aimed to maximise the economic benefits available from ICT and telecommunications to this island, which with its dense population and almost complete lack of domestic resources has always relied upon trade and services to provide for its population. ICT is just the most recent way of trading. More recently, the Singapore government has launched *Singapore One*, an S$82 m project which aims to unite and connect all of Singapore in one broad band network. This will facilitate business and intervention with the government as well as increasing the capacity to boost electronic commerce and also service activities such as teleshopping and telemedicine.

On telemedicine, the European Commission is supporting the *Telematic Health Services for Islands* project which aims to overcome some of the traditional constraints on island health services by such measures as diagnosis and treatment from distance, as well as providing access to information and data on both patients and medicine. The project is being piloted in the Canary Islands (*Insula*, 1998, p. 42). Modern communications technology, for example, enables the dentist on Ascension Island, a Belfastman, to take digital images of cases for which he seeks consultation. He sends these to colleagues in Northern Ireland by e-mail.

Islands, then, can benefit greatly from ICT in both social and commercial fields, and it is no surprise that heavy investment and much interest is taking place. On Scotland's Isle of Skye there is the active Skye Telecommunications Centre which, amongst other things, produces the magazine *Islander* which covers the globe in its discussion of island issues. On Prince Edward Island the university's Institute of Island Studies considers not just PEI but other islands. In 1998 it concluded its North Atlantic Islands Programme and from the Institute is run the Small Islands Information Network, an electronic data system that deals with small islands throughout the world. The International Small Island Studies Association also keeps a watching brief on the ICT field and its fifth triennial conference in 1998 on Mauritius devoted much of its work to ICT and small islands.

Perhaps this section may best be summed up by referring to the *First European Conference on Sustainable Island Development* held in Menorca, one of the Spanish Balearic islands, by the International Scientific Council for Islands Development and UNESCO in 1997. Here was established the *European Island Agenda*, one theme of which was communications and telematics with the realisation that:

> Rapid technological advances in fields like telecommunications and telematic applications offer island economies the potential for real progress throughout the economy, in both public and private sectors. Integrating the islands into the Information Society must be seen as an urgent challenge and a realistic objective that will allow a more rational use of resources and put an end to the traditional isolation of islands.
>
> (*Insula*, 1998, pp. 14–15)

Services

Retailing

The provision of services depends on thresholds. If there is not a sufficient size of population to ensure the profitable operation of a commercial service, it cannot exist unless a subsidy is provided. With regard to education a local authority may be willing for social reasons to keep open a small island school, despite the consequent high per capita cost of the children's education. This type of social service subsidy is usually not available with regard to equally vital but commercial operations, especially retailing. However, it is notable that in Scotland's islands, much of the retailing is carried out by the Co-operative Society, a commercial retailer, but with historical roots in community involvement. On Majuro in the Marshall Islands, as was stated in Chapter 3, there are two multi-functional trading companies, in smaller places there may just be one dominant company, which stifles opportunity for potential entrepreneurs. In the Falkland Islands until the transformations accelerated by the Conflict in 1982, the Falkland Islands Company dominated not just retailing but the whole commercial economy of the islands, landholding, farming, shipping services, almost everything.

Obviously, lower order functions with their smaller threshold requirements are more likely to be provided on small islands than higher order functions. But even the classic examples of daily needs, such as bread and newspapers, are likely to be restricted in supply if the island is too small and remote. Always on an island, except for any domestic production, there is the added cost of all goods reflecting the extra stage of transportation necessary to get the goods to the island in the first place. Further, the normally restricted scale of the small island market means that there are no unit cost reductions to be gained from economies of scale. Additionally, there is unlikely to be full competition and thus prices are not driven down as they would be in larger markets. These factors combine to make small island retailing both restricted in the goods it can offer and expensive. Thus on Ascension Island where there are only twice weekly flights, newspapers appear in the one general store only sporadically and days late. This island's shop is typical in that it is a general store offering a limited choice of relatively high priced goods within a fairly large number of categories, thus fulfilling as many of the customers' needs as possible. Goods not available locally on this or other islands have to be obtained from elsewhere, by visit or by telephone, mail order or the Internet. Travellers to Ascension, mostly arriving on Royal Air Force Tristar jets, may share their plane with fresh vegetables for the military and civilian population of the island, which, apart from fish and a few bunches of bananas, has no domestic food production. There are also scheduled visits from container ships that carry certain food and other supplies. So, one can buy fresh lettuces in the shop in Georgetown, but they cost three times the amount they do in the UK.

Regarding coastal islands, passengers often have to share their ferryboats with boxes of groceries and other goods, which have been ordered from shops in the mainland ports. Or there may be visiting providers of goods and services. Travelling shops, the library van, even travelling banks can be observed in, say, the Western Isles. The idea here is that, by taking the goods around, the supplier is able to reach the threshold population and turnover necessary to obtain a profit. Such travelling services are invaluable to the islanders served but they are available only sporadically, if usually to a schedule, the choice is limited and the goods must be high priced. Inevitably the cost of living on a coastal island is more expensive than in the mainland situation, but at least islanders can usually obtain what they want fairly readily.

In these offshore situations the problems of retail provision are an irritant, but, unlike the provision of some social services like education or health care, restricted on-island retailing opportunities would be unlikely to lead to a decision to migrate. On Ireland's Blasket Islands the people just used to go to the shops in Dingle, just as they had to go to Dingle to attend Mass. On Inish-turk, County Mayo, there is currently no shop, as such, but goods may be bought out of people's kitchens. Islanders adapt, they cope and accept with more or less good grace the problems of retailing as one of the many privations of small island life.

Administrative services

The old adage that nothing is certain in life except death and taxes ensures that every island will be subject to an administrative system that enables taxes to be collected from its inhabitants. Health, educational and judicial services are also subject to administrative control. At the political level, islands are all bound up in a framework of control; even remote, uninhabited places such as Clipperton Island and Bouvet Island belong to some state, even if the administering state is remarkably distant, as Clipperton is from France or Bouvet from Norway. The politics issue will be brought up in Chapter 7. For now we have to consider how day-to-day administrative services are provided in the small island situation. For some offshore dependent islands services may not be provided *in situ* and islanders will have to deal with their business by distance methods or by leaving the island, although the government presence can sometimes be scaled down to provide a minimum level of service. Thus on Grand Manan Island, New Brunswick, Canada, there is a multi-purpose federal government building that provides at least a basic line of communication to the various government agencies and departments. The further up the administrative hierarchy an islander needs to go, the more likely it is that they will have to travel. Thus, for example, a visit to court would probably see people having to leave a small non-independent island. Some may have lower level courts such as a magistrates' court but probably not higher level facilities.

On independent islands things may be different. Here there may well have to be a full availability of a complete administrative hierarchy in such things as justice, in a way that is not necessary even with medicine or education. In independent island states, some patients have to leave their islands for complicated or specialist procedures. Further, as will be discussed below, islands in the Caribbean and the Pacific share the costs of providing university education, which means that those seeking this education may well have to travel off-island for it. This off-island, in fact out-of-state, way of overcoming the limitations of island scale rarely applies to the legal system. Some old British colonies still maintain a link with the British Privy Council, but that apart, independent island countries administer their own complete justice system. Procedures may be inherited from former colonial rulers but they have to be performed within the territory. On Grand Turk, capital island of the Turks and Caicos Islands, a self-administering British colony, the tiny Magistrates' court is next to the tiny Supreme Court. And, like other self-governing territories, the Turks and Caicos Islands has to have a prison. This is on Grand Turk and it is actually a rather fine building (from the outside), one of the most prominent in the miniature city of Cockburn Town (Figure 6.4).

A self-governing island state also has to have the machinery of government, the ministries, civil service, the parliament building. These facilities give some very small urban places levels of administrative function the equivalent of Paris or Rome. Hamilton is the capital of Bermuda, a self-governing dependency of Britain since 1609. It was planned as a replacement to the earlier capital of St Georges, and has upper and lower parliamentary chambers as well as other high order administrative buildings such as tiers of courts and the Cathedral. It also houses a full panoply of ministries and civil service functions. In many island states the scale factor sees ministers doubling up, thus on Bermuda the 13-member cabinet contains a Minister of Works, Engineering and Housing and a Minister of Labour, Home Affairs and Public Safety. On Tuvalu there are 12 ministers including a Minister of Home Affairs and Rural Development and a Minister of Works, Energy and Communications. Autonomous island groups such as Spain's Canary and Balearic Islands do not house national parliaments,

Figure 6.4 H.M. Prison, Cockburn Town, Grand Turk, Turks and Caicos Islands, 1992

but still have to have regional assemblies. Thus in Palma, Mallorca, the Balearic autonomous parliamentary chamber is a few hundred metres from the City Hall in the same way as Bermuda's national parliamentary chamber is a stone's throw from Hamilton City Hall. Such administrative facilities give some capital islands a plethora of functions and jobs whilst the miniature, but fully-fledged cities that result are often quaintly attractive to tourists. So in the circumstances of islands one can find places that lack administrative functions or have only a restricted provision and also islands that have a surprising range of such provisions.

Health and healthcare

The discussion now moves on to health matters. The principal issues are scale and isolation. Sometimes isolation can be a health benefit to an island, for example if a disease does not reach it; thus, in the influenza pandemic after World War I, some island populations were unaffected. However, there are also health difficulties brought about by insularity, especially by the associated isolation (Cliff and Haggett, 1995). In a situation where there is a constant movement of people and their ailments, tolerance can build up within a society towards infectious diseases. Measles, mumps, chickenpox – such diseases became irritants or childhood problems, of major risk to only an unfortunate few, or people weakened by age and debility. However, in situations where, because of isolation, tolerance through constant exposure cannot build up, what are minor diseases elsewhere can become a problem if introduced. Island peoples by definition are isolated, less so now of course, but certainly they were often very isolated in the past. Contact with outsiders carrying disease could be devastating to a population not hitherto exposed to this disease because of their isolation. Thus island populations were very prone to epidemics. This was appreciated a long time ago. A ship's surgeon, William Gunn, visiting Pitcairn aboard *H.M.S. Curaçoa* in 1841, was dismissive of an external cause for an influenza epidemic he found there, but wrote of his fears for the introduction of other diseases as well as harm to the island society being caused by contact with outsiders:

> Although they have as yet remained exempt from contagious diseases, it is not probable that they will continue so – their rapidly-increasing communication with ships is not only likely to give origin to these and other complaints, now unknown amongst them, but also to produce a change in their present simple method of living as well as in their moral conduct.
>
> Boils and eruptions are not uncommon amongst them now, and are most probably caused by the change produced in their diet pending the visit of ships. Their traffic with these vessels, consisting chiefly of barter for old clothes, greatly increases the risk of that dreadful pestilence, the small pox, being soon imported amongst them.
>
> (reproduced in Royle, 2000b)

Smallpox was Surgeon Gunn's particular worry but isolated groups could be decimated by much milder diseases, say, measles. Such groups did not have to be islanders of course. Native Americans, in both North and South America, were hit by disease brought in by early European explorers – the retaliation was certain sexually transmitted diseases which were taken back to Europe. On islands, however, the situation could be even worse, given the greater isolation of the peoples. Few aboriginal populations survived European contact in the West Indies. The Beothuks were wiped out in Newfoundland; nor did the aboriginal population of Tasmania survive; though some were deliberately killed rather than dying of disease.

Off Canada's west coast, the Haida are the First Nation people of the Queen Charlotte Islands (named for the ship of British sailor, George Dixon, in 1787, but known as Haida Gwai to the First Nation). In 1840 a census recorded 6693 people living in 12 main villages. Sadly European contact brought smallpox to the islands, as well as venereal disease and tuberculosis, conditions against which there was no resistance amongst the Haida. By the 1850s thousands of Haida had died and others had left the islands. At the end of the nineteenth century there were only about 600 Haida left, living in just two villages (Dalzell, 1968). The abandoned village of Ninstints to the south of Moresby Island had a particularly fine expression of Haida culture with wonderful totem poles. It is now a UNESCO World Heritage Site – mute tribute to the inability of its people to protect themselves against the Europeans, not necessarily against their weapons, but against their diseases. Decimation of insular populations by introduced diseases is now, thankfully, a thing of the past, but there can still be problems. Thus St Helena, accessible only by ship, is often subject to introduced epidemics, in 1986 all schools had to be closed after a flu-like viral infection was brought in on the island's ship (Bain, 1993).

Another health problem associated with insular isolation is the predisposition of such populations to establish endemic diseases and conditions. Oliver Sacks' curiously titled book, *The island of the colour-blind and Cycad Island*, emanated from his journeys to Micronesia as a

> neurologist, or a neuroanthropologist, intent on seeing how individuals and communities responded to unusual endemic conditions – a hereditary total colour-blindness in Pingelap and Pohnpei; a progressive, fatal neurodegenerative disorder in Guam and Rota.
>
> (1996, p. xii)

These types of shared conditions might be associated with a shared exposure to environmental factors or have something to do with a restricted gene pool. For example, the population of Tristan da Cunha has a high incidence of asthma. One third of the people are affected. What is it about Tristan or Tristanians that causes them to develop this disease? If this can be discovered then this better understanding of the disease might help in the development of preventative or curative measures. An American company is now investigating this population to try to find out (Coghlan, 1996; Zamel, 1995). Similarly, the population of Iceland is being used for disease studies. Here is an isolated group of people, inter-related with a restricted gene pool. Most importantly, Iceland has good health and genealogical statistics, dating back to the *Book of Settlements* of 1130, which lists the principal original settlers and their genealogies back to their own roots in Norway, the British Isles and elsewhere. This material may enable genetic predisposition to certain diseases to be identified (Hannibalsson, 1999).

With the exception of some endemic disease predisposition, the health problems of most islands in the modern era are not particularly different from those of their region. Islands sit within their region and share the health problems of their region. Malaria was a significant problem on Mauritius as in other parts of tropical Africa; inhabitants of the Scottish islands tend to have a similar diet and thus similar dietary-related health problems as citizens of Glasgow. The particular problems of diet-related diabetes associated with the problems of urbanisation and modernisation in the Marshall Islands were highlighted in Chapter 5. These problems are shared with other Pacific islands, for example, in Truk State (Cameron, 1992); but in others, such as Kiribati, the major health problems are those of the modern world: respiratory diseases, diarrhoea and heart and circulation problems. The major cause of death in I-Kiribati adults is cardio-vascular disease. What remains an inescapable problem for health matters on islands in the present as well as in the past, and one that is caused by insularity, is health care delivery, as the next section will discuss.

Health care delivery in an island setting

It is the duty of the nation state to which an island belongs, to provide a health care system appropriate for that state in the island setting. Islanders in offshore parts of Italy and Spain would expect to receive access to health care facilities of similar order to citizens of Madrid or Rome; offshore residents of, say, Equatorial Guinea would have much lower expectations, as would their mainland fellow nationals.

The trick for island health care systems is how to deliver this appropriate level of care in the insular setting. If funds are not an issue, island health care can be good. In the French overseas island territories in the period since World War II health care has been considerably improved, as might be expected of these parts of France. Matters like malaria control could and did see huge reductions in mortality. In consequence, the death rate in Réunion fell from 19 per thousand in 1957 to six per thousand in 1981 and life expectancy rose from 50.5 years in 1959 to 70.5 years in 1984, a figure equivalent to that of metropolitan France. Further,

> in Martinique . . . the number of doctors more than doubled from 1961 to 1979 and infant mortality dropped by half; in Réunion the number of physicians similarly grew almost fifty-fold and the infant mortality rate fell by nine-tenths.
>
> (Aldrich and Connell, 1992, p. 147)

In New Caledonia, France also provides an efficient and modern health service. The main causes of death there are now similar to those found in the modern temperate industrialised world.

Réunion and New Caledonia have populations in the hundreds of thousands and thus have the scale as well as the finances to sustain a full health service, although when necessary, patients are air lifted to Australia for treatment from New Caledonia. Smaller islands have more problems, even when they are part of the developed world. Small populations do not generate sufficient demand for medical care to make its provision cost effective. Even where there may be an island population of sufficient size for a doctor, even a hospital, there may well not be enough demand for specialist facilities to be provided on the island. Health service administrators are not keen to provide staff who would be under-employed or machines that would be idle for a large part of the time. Therefore island patients requiring many types of treatment may have to leave, or if necessary, be evacuated, to obtain the care they need elsewhere. This procedure is both costly – the person or the health care system has to bear the cost of transportation, perhaps also of accommodation – and inconvenient, at least to the patient. It may also be risky to the patient, depending on the type of condition and its severity – journeys by ship or plane are hardly therapeutic to the seriously ill. Tiny islands may not have any medical provision at all, with even a visit to the doctor requiring travel off the island. Alternatively, peripatetic medical help may be provided. On Tory Island, County Donegal, the general practitioner calls by helicopter in winter when the island can be cut off from sea access for days, even weeks, at a time by storms. A nurse lives on the island.

In extremis, emergency medical evacuation (medevac) may have to be carried out. Thus all the inhabited Irish islands now have helicopter pads, and helicopters on call for coastal and shipping emergencies can be scrambled to the islands in a relatively short time. The author witnessed a very efficient helicopter medevac of a building worker who had fallen from a roof on Clare Island, County Mayo in 1997, for example (Figure 6.5).

Islands in the developing world make similar types of arrangement as appropriate to their financial resources. Thus Kiribati has a hospital on South Tarawa but it cannot deal with orthopaedic surgery, uncommon surgical procedures and serious cardiac operations, and such procedures have to be carried out overseas. Residents of the off-islands would have local clinics, but may well have to visit South Tarawa for even fairly routine procedures. However, there is also a ship provided by an international charity that takes medical care to the outer islands. This supplements the health care the Government is able to provide in and from South Tarawa. The Kiribati situation points up some of the difficulties in providing an equivalence of health care to patients in the capital and off-islands. Thus, on the Falkland Islands, there is a good new hospital in Stanley but this is difficult to reach from the farm settlements in the rural areas and off islands. The Government's medical department has to strive to take care to these settlements; thus, in 1998, a dental surgery was established in Fox Bay, one of the West Falkland island farms. This 'should significantly improve the

Figure 6.5 A medevac from Clare Island, County Mayo, Ireland, 1997

dental care available to the residents of West Falkland' (Falkland Islands Government, 1998, p. 16; see also Royle, 1995).

Education

Education cannot be divorced from other social processes, so islands experiencing major in-migration, such as the capital islands mentioned in Chapter 5, will find that their educational system is under as much pressure as anything else. On Majuro in the Marshall Islands, for example, pupils have to attend in shifts; such is the pressure on facilities. In other island situations, the problem is in having to cope not with too many pupils, but with a small number in a remote place. The level of educational provision available to an island's population depends to a great extent upon national norms, but education and its provision are also sometimes directly affected by a place being an island.

The most obvious problem relates, again, to scale. The most basic level of education is that of the primary school. An island without sufficient people for a primary school must be in serious difficulties regarding its continued habitation in even the short term. Parents are very reluctant to let children not yet into double figures be sent from them to be educated in a boarding situation elsewhere. Only marginally less unacceptable would be to subject young children to ferry journeys twice a day to a place where education was provided, even if this were possible. Thus, once a decision is taken to close an island's primary school, the death knell of an island is sounded, and almost inevitably parents with school

Figure 6.6 Inishbofin National School, 1899–1981, Inishbofin, County Donegal, Ireland, 1984

age children will leave. If the island school closes because there are no school age children anyway, then the island is already dying. Off Ireland, the school on Inishbofin, County Donegal, a National School dating from 1899, closed in 1981 and the island is now only seasonally populated (Figure 6.6). The primary school on Caldey Island off South Wales closed in 2000 after more than 100 years, when two of the four pupils left. The headteacher recognised that 'without a school the island is unlikely to attract families with children' (*Independent*, 21 July 2000). Islanders and their governing authorities are naturally aware of the importance of primary education in the island setting. So where it is policy to support island people *in situ*, one factor of such support must be the continued provision of primary education, even on a very expensive per capita basis.

The author was once on Sherkin Island, County Cork at school break time and observed the pupils lining up to return to class, the range of sizes typical of a primary school represented amongst the six children. In Northern Ireland's only inhabited offshore island, Rathlin, there is a primary school provided within the Province's Catholic sector educational system, all the islanders being Roman Catholics. In the 1980s one family refused to send their children to this school after a dispute and exercised their right to have their children educated in the parallel state sector educational system. The local education authority was unable to insist that the family send their children to school in Rathlin's mainland port of Ballycastle. Instead, the board had to deliver to the island two portacabins, one to serve as the family's school, the other to act as a house for the teacher. The cost was considerable.

In the Falkland Islands there is a need for primary education to be provided in Camp, the rural areas. In the late nineteenth century there was concern that proper education was not being provided for all even in Stanley, whilst 'the state of education in the country parts of the islands is deplorable' according to the Inspector of Schools in 1887 (see Evans, 1994, p. 323). Only one Camp settlement had, at that point, a proper school; elsewhere children were being taught by their parents, people who, in many cases, were ill educated themselves. The schools inspector, Reverend Lowther Brandon, interestingly suggested that enquiries should be made to see how 'other island communities in the world coped with similar difficulties' (Evans, 1994, p. 324). Only in 1895 was education made compulsory across the islands and the Falkland Islands government began to employ peripatetic teachers for Camp, at £5 per month, plus free passage to the Falklands. All Camp settlements were and are farms, and they are widely scattered around the two large islands of East and West Falkland and on the many off-islands that are managed as single farms. The peripatetic teachers travelled to those farm settlements that had primary aged children and stayed for two or more weeks, receiving free board and lodging in exchange for the teaching. In the days when the only mode of travel across the Falklands was by horseback, some small and remote settlements may have received only one visit per year and in 1905 it was admitted that children on two very inaccessible off-islands, Speedwell and Bleaker, were not being visited. On their departure from a farm settlement, the teachers left work and gave instructions to the parents who were then responsible for seeing it completed, although it was widely acknowledged that in some cases it was a forlorn hope that the education would be continued properly. Today there is a Camp Education Unit, which provides for the education of rural children of primary age. Peripatetic teachers are still employed. In 1997 extra staff were taken on to reduce the periodicity of the 'beats' to six weeks from eight. With the improvement in intra-island communications, farms often send their children to a neighbour's farm when the teacher is there. Camp settlements usually have a dedicated schoolhouse. The Camp educational system also makes use of radio and other telecommunication technologies to continue the education of the children during those periods when the peripatetic teachers are elsewhere, and an additional radio/telephone teacher was also recruited in 1997. Three times a year, the Camp Education Unit produces *Classroom Wall*, a booklet which enables these isolated Camp children to display their work to each other. During Farmers' Weeks, when Camp people traditionally come into Stanley, many bring their children and they get the experience for a while of attending school in a group. Special projects and activities are set up for them (Falkland Islands Government, 1998).

Secondary education needs to gather together pupils in reasonable numbers for it to be suitably provided. Parents on small islands normally accept this, and it is usual for children from such backgrounds to have to go elsewhere and board when they reach the age for secondary schooling. Thus few of the off-shore Irish islands have secondary schools. In the Falklands the one secondary school, the Falkland Islands Community School, is in Stanley and children from

Camp board in the hostel accommodation provided. For A level and university education, the Falklands government pays for students to attend schools and colleges in the UK.

On Ascension Island there is one school. Ascension is so small and remote that it is not possible easily or cheaply to provide post-primary education off-island. So Two Boats School, with 97 pupils in 1999–2000, provides education and public examination facilities up to the age of 16, which takes children through to their British GCSE examinations, this being a British Overseas Territory. There are nine teachers and so class sizes are small. The children benefit from much individual attention, but teachers report that group and discussion work can be difficult. Any pupils wishing to go on to sixth form to take A Levels have to leave the island; in 1999–2000 there were two studying in England.

Tertiary level education also normally requires movement off a small island. In large island nations such as the Philippines or Indonesia there are universities and colleges on different larger islands. In other island situations there may be sufficient scale of population to support only a single college or university. There is one university on Mallorca, for example. The small University of Prince Edward Island (enrolment c. 2900 with 200 faculty) serves the Canadian province (population 134,000). Rather smaller is the University of the Faroe Islands (population 50,000), which has just 16 faculty. On Cyprus in recent years, university education has been provided and, given the island's division, this has necessitated facilities in both parts of the island. In the south, the University of Cyprus took in its first students in 1992. Presently housed in buildings in Nicosia, a purpose built campus is under construction at Athalassa, outside the capital. In the north, The Eastern Mediterranean University began as a Higher Technical Institute in 1979, before receiving its university charter in 1986. Its campus is in Gazimagusa (Famagusta) and the Turkish Higher Education Council accredits its courses.

On occasion, however, there is just not sufficient mass within an island nation for full tertiary educational opportunities to be provided. Then scaling-up operations may have to be set up. Two fine examples are the University of the West Indies (UWI) and the University of the South Pacific (USP). UWI serves the islands of the English-speaking Caribbean and Guyana on the Latin American mainland. A similar arrangement pertains with USP regarding the English-speaking Pacific islands. Twelve nations: Cook Islands, Fiji Islands, Kiribati, Marshall Islands, Nauru, Niue, Solomon Islands, Tokelau, Tonga, Tuvalu, Vanuatu and Western Samoa are constituents of USP and are eligible to send students to the main campus. This is in the area's largest population centre, the Fijian capital city of Suva on the island of Viti Levu. USP maintains out-stations on other islands and some students can carry out work through out-stations and distance education techniques. USP staff members travel extensively throughout the region to provide tutorial support.

Distance education is also part of the strategy of the recently established (1998) University of the Highlands and Islands in the UK. It is a collegiate university that has campuses in 13 locations on the Scottish mainland and islands, including facilities in Shetland, Orkney, Lewis and Skye. In addition it has satellite learning centres throughout the Highlands and Islands region. All its facilities have high quality telecommunications links to facilitate distance learning.

The University of the Highlands and Islands offers standard university courses, but also some specifically designed for the rural and island situation of its area. These include Environmental and Heritage Studies, Rural Development Studies, Marine Science and also courses in and on Gaelic, the ancient language now spoken largely in the Western Isles and Skye. This introduces another issue regarding island education – its appropriateness. Distant colonial islands may have been saddled with subjects and syllabuses more appropriate to their continental metropoles than the island itself. In the modern era there is more sensitivity to the local situation, thus the one school on Tristan da Cunha takes pupils to a limited range of GCSE examinations in subjects including English, mathematics and Tristan Studies. At the tertiary level, the one university on Cape Breton Island, Canada, University College of Cape Breton, offers a number of courses that work with and focus on the local community. It also participates with universities in Iceland, Scotland and Northern Ireland in a North Atlantic Studies programme. There is a realisation though, that the local economy being weak, many islanders seek education only to increase their life chances upon migration after graduation.

In sum, this chapter has demonstrated that regarding education and other services, provision is always more problematic in an island situation, although there have been developed a number of strategies to try to ease the situation. It has also shown that transportation is of vital interest to small islands. Only with regard to telecommunications are such islands sometimes less troubled by their isolation.

References

Key readings

Cliff and Haggett write briefly on the epidemiological significance of islands, whilst Sacks's book details some of the impact of isolation on health status. Evans's book is a detailed study on educational provision in a series of different small island settings. Hamley writes of the impact a major new transportation investment can make on a small island; Kokovkin writes on the impact of telecommunications.

Aalen, F.H. and Brody, H. (1969) *Gola: the life and last days of an island community*, Mercier: Cork.

Aldrich, R. and Connell, J. (1992) *France's overseas frontier: Départements et Territoires d'Outre-Mer*, Cambridge University Press: Cambridge.

Bain, K. (1993) *St Helena: the island, her people and their ship*, Wilton 65: York.

Begley, L. (ed.) (1993) *Crossing that bridge: a critical look at the PEI fixed link*, Ragweed Press: Charlottetown.

Cameron, J. (1992) 'The Federated States of Micronesia: is there a Pacific way to avoid a MIRAB society?', in Hintjens, H.M. and Newitt, M.D.D. (eds) (1992) *The political economy of small tropical islands: the importance of being small*, University of Exeter Press: Exeter, pp. 150–79.

Cliff, A.D. and Haggett, P. (1995) 'The epidemiological significance of islands', *Health and Place*, 1, 4, pp. 199–209.

Coghlan, A. (1996) 'Gene treaty promises rewards for unique peoples', *New Scientist*, 2 November, p. 8.

Dalzell, K.E. (1968) *The Queen Charlotte Islands, 1774–1966*, Bill Ellis: Queen Charlotte City.

Didier-Hache, J. (1987) 'The politics of island transport', *Scottish Government Yearbook*, 1987 Edition, pp. 124–42.

Donne, J. (1624) *Devotions vpon emergent occasions, and seuerall steps in my sicknes*, Thomas Iones: London.

Evans, D. (1994) *Schooling in the South Atlantic islands, 1661–1992*, Anthony Nelson: Oswestry.

Falkland Islands Government (1998) *Report of the Governor on the financial year July 1997 to June 1998*, Falkland Islands Government: Stanley.

Hamley, W. (1998) 'The Confederation Bridge and its likely impact on the economy of Prince Edward Island', *British Journal of Canadian Studies*, 13, 1, pp. 140–7.

Hannibalsson, J.B. (1999) 'The raison d'être of the Icelanders', *Iceland Review*, February 1999, pp. 28–33.

Kokovkin, T. (1998) 'Hiiumaa: an island on the internet', *Insula*, 7, 1, pp. 17–18.

Round Table on Resource Land Use and Stewardship (1997) *Cultivating island solutions*, Round Table on Resource Land Use and Stewardship: Charlottetown.

Royle, S.A. (1995) 'Health in small island communities: the UK's South Atlantic colonies', *Health and Place*, 1, 4, pp. 257–64.

Royle, S.A. (2000a) 'Community development in a restricted ecumene: the case of Prince Edward Island', *British Journal of Canadian Studies*, forthcoming.

Royle, S.A. (2000b) 'Health and health care in Pitcairn Island in 1841: the report of Surgeon Gunn', *Journal of Pacific History*, 35, 2, pp. 213–17.

Royle, S.A. and Scott, D. (1996) 'Accessibility and the Irish islands', *Geography*, 81, 2, 1996, pp. 111–19.

Sacks, O. (1996) *The island of the colour-blind and Cycad Island*, Picador: London.

Zamel, N. (1995) 'In search of the genes of asthma on the island of Tristan da Cunha', *Canadian Respiratory Journal*, 2, 1, pp. 18–22.

7   Politics and small islands

Island powerlessness

Some of the world's largest nations are insular. Indonesia, at around 192 million people, has fewer inhabitants than only China, India and the USA. Japan (c. 123 m) is the seventh largest nation in the world; the Philippines (c. 64 m) comes 14th and the United Kingdom (c. 58 m), 16th. In terms of area, Indonesia at 1.9 million km^2 is the world's 15th largest country; Greenland, at almost 2.2 million km^2 is even bigger, in 13th place. On occasion, some of these island nations have had economic and/or political power commensurate with their size. The United Kingdom entered the twentieth century as the world's leading power and controlled one-fifth of the world's land surface and a quarter of its population. Japan, too, had an extensive regional empire in Asia, stretching into Australasia on occasion, during the twentieth century. Upon recovery from its eventual defeat in World War II, Japan, like Germany, made a remarkably quick recovery and ended the twentieth century as one of the world's mightiest economies.

That there have been large and powerful island nations does not gainsay the fact that the more usual insular position regarding politics, especially for small islands, is one of powerlessness, dependency and insignificance. The problems of scale, isolation, peripherality, etc., normally handicap small islands in the political arena as they do in every other aspect of human life. An outside force can usually scale up sufficient economic, political and/or if necessary, military force to impose their will upon the people of an island. If islanders resist what is proposed for them, history, including very recent history, gives us scores of examples of islands being invaded. When it comes to war, small islands are very vulnerable. Thus the only parts of British homeland territory to be taken by Germany during World War II were the Channel Islands from which the British forces withdrew, recognising that the islands could not be held:

> They were wide open to attack from France by sea and air. To defend them adequately would be costly. To defend them inadequately would expose the people to the horror and privation of war for no good reason.
>
> (Cruikshank, 1975, p. 23)

King George VI's message to the Bailiffs and thus the islanders spoke of the strategic necessity of abandoning the islands and that it was 'in the islanders' interest that this step should be taken in the present circumstances' (cited in Cruikshank, 1975, p. 32). By contrast, the British took control of other friendly European islands rather further from German bases in the hope that they could keep them from German hands. These were Iceland and the Faroe Islands. The Americans took Greenland for similar reasons.

If there is actual conflict and island defenders lose control of the sea, or, in recent years, the air, then they are forced back upon domestic defences and resources and these are usually insufficient to beat off an attacking force which has mustered sufficient supplies and soldiery. Thus, even the powerful and determined Japanese were forced to surrender Okinawa in the face of over-whelming American military superiority in 1945. The fact that there are rare occasions when a determined force has failed to conquer an island is perhaps just the exception that proves the rule. The best example of such an exception relates to Malta and the Great Siege of 1565 (see Chapter 3).

The case of the Falkland Islands

There are countless other examples of islands being taken by invasion. Such stories are not only from times long ago. In 1673 the Dutch expelled the British East India Company from St Helena, only to be expelled in their turn within a few months; but then 309 years later, in 1982, in the same ocean came the Falklands Conflict, when history basically repeated itself.

The Falkland Islands were first definitively discovered by the British; truly discovered, for they were uninhabited. This was by the *Desire*, Captain John Davis, in 1591, although there were possibly earlier sightings (see Goebel, 1927). No attempt was then made to land or settle. The British were probably the first to land, too, in 1689. But the French were first to settle, at Port Louis on East Falkland in 1764 in a private venture under nobleman, Louis-Antoinne de Bougainville. The party claimed what they called Les Malouines (after St Malo) for France and in the first structure built, a small fort, an obelisk with the king's head was erected (Philpott, 1996). The Spanish objected to this settlement on an island that they regarded as an offshore part of their South American territories. The French government was unwilling to dispute the matter and pressure was put on de Bougainville to relinquish his colony. In April 1767 he sold his settlement and rights to the Spanish, and most, but not all, the French settlers left. Meanwhile, independently, the British had finally settled on the Falklands, in 1766, founding Port Egmont on Saunders Island off West Falkland. After the British and Spanish discovered each other's presence, the garrison at Port Egmont lost a skirmish to the Spanish from Buenos Aires in 1770. It seemed that war might break out between Britain and Spain in consequence. There was not war in the event, but much diplomatic intrigue, and in 1771 Port Egmont was restored to the British. In 1774, the British abandoned the islands, ostensibly as a cost-saving measure, although Goebel opined that this was the result of an 'antecedent obligation' (1927, p. 409), in other words a face-saving deal had been done. The British left a plaque at Port Egmont to proclaim that the Falkland Islands were 'the sole right and property of His Most Sacred Majesty George the Third, King of Great Britain, France and Ireland'. The United Provinces (Argentina) assumed the Spanish claim, upon that country's declaration of independence in 1816. The Spanish had previously abandoned the Falklands in 1811 and the islands were then without governance, although used fairly regularly, by whalers especially. In 1820, keen to secure all territory once ruled by Spain, the United Provinces sent a party to reclaim and resettle the Falklands, later using feral cattle, gone wild from earlier ventures, as one of the resources. There were often disputes between the new settlers and foreign sealers, and in 1831 three American ships were seized. The Americans sent a frigate, the *Lexington*, in a punitive raid in response. The habitations were tumbled; the guns were spiked, almost all the inhabitants expelled and the islands once more declared free of all government. The Argentines returned in 1832 and there was violence with a Briton, who had been on the island from before 1831, being killed. At this point the British navy, in the form of *H.M.S. Clio* came in to restore order and the British took the islands and the Argentines were expelled. Their government protested and never abandoned their claim (Goebel, 1927).

In January 1833, when the *Clio* took the islands, there were just nine settlers left, and they then had to be supported by the issue of British naval rations. Nine years later the British Lieutenant-Governor, Richard Moody, was ordered, against his will, to move the population from Port Louis to what became Stanley and he laid out the grid plan settlement that remains the islands' only town. The economy became focused upon ship revictualling, dealing with vessels rounding Cape Horn; and later the territory was opened up for extensive sheep ranching and that became the islands' principal trade for many decades (Royle, 1985). The Falklands had a role in naval engagements in both World War I (the Battle of the Falklands) and World War II (the Battle of the River Plate), but were largely unknown to the world when they were forced into global headlines in 1982. The Argentine claim to the Falklands, to them called the Malvinas (a name rooted to the original French name) had never been abandoned. In the 1970s the British investigated the possibility of decolonialising the islands, which would have been against the will of the people. However, in 1982, the Argentine leader, General Leopold Galtieri, impatient at the slow pace of political change and also needing to stimulate domestic political unity, ordered an invasion. There were British troops on the islands, but only a token force, and they and the local Falkland Islands Defence Force, a tiny militia, were completely unable to resist an invasion at Stanley. To the surprise of much of the world, the UK government under the leadership of Margaret Thatcher, promptly organised a task force to relieve the islands and they became British again (Figure 7.1). This determination by the British Prime Minister was as much for domestic political consumption as had been the original invasion. The 'Falklands Factor' was seen as a large positive element in the victory of

Figure 7.1 The British Governor at the Queen's Birthday celebrations, Stanley, Falkland
Islands, 1993

Thatcher's Conservative Party at the British general election in 1983. The
British gained control of the sea, the controversial sinking of the Argentine
warship, *General Belgrano*, was a factor here, and later the air, although the
British first lost a number of ships to Argentinian aircraft. The Argentinian
forces on the islands focused their defences at Stanley; the British, in a move
reminiscent of the Moslem invasion of Malta 417 years earlier, invaded instead
at a less defended spot and made their way overland to meet the enemy forces.
Once the Argentines on the islands were cut off, the usual victory to the invading forces took place (see, for example, Strange, 1985).

   Two points emerge from histories like that of the Falklands over the centuries. Firstly, that islands are difficult to defend against invasion, especially
when they are cut off from succour from outside, and secondly, that islanders
are often themselves of little regard in the playing out of political games involving external powers.

Internal and external colonialism

The point above is reinforced by the way in which islands were caught up in
colonialism as Chapter 3 demonstrated. Sometimes the colonialism was/is
internal, for within island groups there can be quasi-colonial power relationships, just as there are between islands and continental areas. Within the British
Isles, Ireland, the smaller of the two main islands, was dominated by its larger
neighbour for centuries before part of it gained political independence in 1921.

Economic independence only came much later. Ireland had perforce to join what became the European Union in 1973 along with the UK, its then over-whelmingly dominant trading partner and to whose currency its own was fixed. Only with Ireland's decision to join the European Union's common currency, the Euro, from the start in 1999, whilst the UK remained outside, was the smaller island nation's economic independence truly demonstrated and symbol-ised.

Internal colonisation might also be recognised in Japan's assumption of control over Okinawa and the other Ryukyu Islands to its south in 1879. Okinawa is the one part of Japan that still houses a considerable foreign military presence, with one-fifth of the island being given over to US bases (see below). Some commentators have seen this continuing foreign dominance as a sign of disrespect by the Japanese for this, the poorest and most distant part of the country and the only area with a non-Japanese culture (Desmond, 1995b).

Thus, within Japan and the British Isles one can recognise internal colonial-ism directed from the dominant core against their weaker island peripheries. However, even these powerful island nations join all island nations in the world in not always having been able to maintain their own freedom from external political control. Even the four large island nations spent some time being ruled by outside groups. Indonesia was a colony of the Netherlands as the Dutch East Indies and became independent in 1947 in the wave of decolonisation from the exhausted European powers that took place at the end of World War II. The Philippines have had experience of belonging to more than one power, having been Spanish from 1565 until being handed over to the United States in 1898. Independence came in 1946, but the Americans retained a quasi-colonial pres-ence on the islands until finally abandoning the huge military base at Subic Bay in 1992. England, to deconstruct the not really United Kingdom, has been self-governing for many centuries but was settled and ruled by a range of conti-nental European powers before that, including the Romans, the Vikings, the Anglo-Saxons, and, most recently, the Normans. The fact that 'most recently' refers to the invasion of 1066 does not negate the fact that even England has not always been independent and free from foreign control. The Welsh, Irish and Scots have been dominated by England and also have had their share of invasions, although the Romans did not rule either Ireland or Scotland; Hadrian's Wall still testifying to the presence of the Roman boundary across the island of Great Britain.

Japan has had as nearly a free history as any insular nation can claim but even here there was a recent loss of political control when defeat in World War II led to the islands being administered by the Americans for a short time. Parts of Japan, including Okinawa, were not handed back until 1972.

Many coastal islands are simply part of the neighbouring mainland nation and have shared its political history. It is in the nature of islands, however, that sometimes there has been a different journey, if the island has been used as a strategic base by another power or as a stepping stone as explained in Chapter 4. Thus Zanzibar has not always been linked politically with the rest of Tanzania; the Balearic Islands have not always been politically united with mainland Spain. Not that these places were always self-governing before: Zanzibar was an Arab trading base and belonged to Germany and then Britain. The Balearics had a period of self-rule in the Middle Ages but were, like other Mediterranean islands, pawns of outside rulers at other times; the Moors and the Romans took them and the legacy of these groups can still be picked out in the landscapes of Mallorca. On neighbouring Menorca, the presence of a splendid harbour saw this island also passed between the British and the French. Now, despite various degrees of autonomy, these and other coastal islands belong to the nation state across the water – the Spanish Prime Minister signs international treaties on behalf of the Balearics as much as on behalf of Madrid. Usually coastal islands belong to the nearest mainland power. There are some exceptions; the results of centuries of enmity and competition between Greece and Turkey has seen many islands within sight of the Turkish coast actually being part of Greece; Bornholm belongs to Denmark, not to the nearer state of Sweden.

Many oceanic islands are now independent, but all spent some time as at the least protectorates of colonial powers; most were full colonies; some still retain colonial status. Thus, Fiji, now independent as the Fiji Islands, was ceded by King Cakobau to Queen Victoria's Britain in 1874. Accessible by Air Marshall from Fiji are Tuvalu and Kiribati, now separately independent but once British together as the Gilbert and Ellice Islands Colony. The Air Marshall plane then carries on to the Marshall Islands which had the dubious pleasure of being German and Japanese before being put in the care of the Americans, from whose Trust Territory they have recently emerged. Southeast from Fiji, flights go to Tonga, never a colony *per se* but a British protectorate. East from Fiji, Air New Zealand flies to the Cook Islands which are still, like neighbouring Tokelau, administered by New Zealand, and onward to French Polynesia, still French. Fly on to New Zealand itself, formerly British. Fly northwest from Fiji on Air Nauru to Nauru itself, now independent, formerly ruled jointly by the UK, Australia and New Zealand in the form of the British Phosphate Commissioners, although only the two Pacific countries used the phosphate. This was under a League of Nations mandate, for from 1888 Nauru had been German. West from Fiji is Vanuatu, once administered jointly as the Anglo-French Condominium of the New Hebrides. Papua New Guinea has had a complex political history involving Germany, Australia and Britain before becoming independent. Nearby New Caledonia is still French, as are the Wallis and Fortuna Islands to Fiji's northeast. Next to them are the Samoan Islands. Samoa was once German and then administered by New Zealand before independence; American Samoa is still American, as is Guam and the Northern Mariana Islands; the Federated States of Micronesia and Palau came out of the American Trust Territory. Isolated islands belong to Australia (Lord Howe and Norfolk) or to New Zealand (the Kermadec and Chatham groups) or to the USA (Johnson, Wake (disputed by the Marshall Islands), Howland and Baker). Further north is the Hawaiian group, a state of the USA since 1959, before that a US Territory, taken by force by American marines in 1893. There is still a current British Pacific presence in the form of Pitcairn, the world's smallest separately administered territory with around 50 people. East from Pitcairn start a series of islands such as Easter Island (Rapa Nui) and Sala y Gomes administered by Chile.

In the Atlantic there is a whole series of islands starting from Iceland and working southwards that have been produced by the mid-Atlantic ridge. Iceland, though tracing its parliamentary history back to the Althingi in AD 930 has been independent in the modern era only since 1944, having been both Norwegian and Danish and taken under the control of the British during World War II. To the north Svalbard (Spitzbergen) is Norwegian, with also a Russian presence; Bear Island and Jan Mayen are Norwegian; Franz Josef Land, Russian; south of Iceland, Rockall is, as we saw in Chapter 1, disputed but still owned by somebody. The Azores are part of Portugal; St Paul's Rocks are Brazilian; a string of volcanic islands on or now dragged away from the ridge, from Ascension through St Helena, to the Tristan da Cunha group including Nightingale and Gough are British; further down the ridge, before Antarctica is reached, is uninhabited Bouvet Island which belongs to Norway. West of Bouvet are uninhabited islands such as South Georgia, the South Sandwich and South Orkney groups, formerly used, especially in the case of South Georgia, by a number of whaling nations but now administered by the UK, as are the Falkland Islands, if still subject to Argentina's claim. On the Latin American side of the Atlantic ridge are a number of Brazilian islands: Trinidade, Martin Vaz and the Fernando de Noronha group. Leaving aside the Caribbean, further north off America come the Bahamas, formerly British, and the Turks and Caicos, Atlantic not Caribbean islands strictly speaking, which are still British, as is Bermuda. There are strings of American islands along its east coast, the most northerly of which, Machias Seal Island is disputed by Canada. Canada has its own Atlantic islands, some of which, such as Cape Breton, spent time in French as well as British hands before becoming part of an independent Canada. The largest Atlantic Canadian island is Newfoundland, which remained British until 1949. Further north, the largest island of all, Greenland, was Danish and although now self-governing, remains part of the Danish realm, a situation shared by the Faroe Islands, situated between Scotland and Iceland, although there are pressures here for independence. On the African side of the Atlantic are some few coastal islands, including South Africa's notorious Robben Island, once the prison of Nelson Mandela, amongst others. There are more islands in the Gulf of Guinea, such as Annobon and Bioko (Fernando Po), once Spanish and now part of Equatorial Guinea. São Tomé e Príncipe, and the Cape Verde Islands have moved to independence from Portugal; around the coast are Spain's Canary Islands, and Madeira, still Portuguese. The political status of the British Isles has already been discussed.

In the Mediterranean there are only two independent islands, Malta and Cyprus; both were occupied and controlled by a multiplicity of different foreign powers, those ruling Cyprus were listed in Chapter 3. Northern Cyprus is still under the effective control of Turkey, following the invasion of 1974.

A similar exercise could be carried out in those parts of the Pacific not considered above, for the Caribbean and for the Indian Ocean and its offshoots such as the Persian Gulf. In these maritime areas there is also a mixture of dependent coastal islands, independent states, which used to be colonies or at the very least, like Bahrain, protectorates; and islands that are still colonies. In sum, the political map of the world's island realm is very complex; the political history even more complex, the one common factor is that all islands in the past and many still in the present were/are under the control of an outside power: island powerlessness.

Islands, foreign domination and dependency

War is not the only way in which the strong control and manipulate the weak. Islands, having been subject to colonialism, are often subject to the foreign domination that is neo-colonialism. Islanders can be outvoted; island economies can be bought up. Whole islands have been purchased, thus Denmark sold three of the Virgin Islands – St Thomas, St John and St Croix – to the USA in 1917, ridding their state of a financial burden, the Americans gaining a strategic asset. Islands have even been swapped, thus in the Zanzibar Agreement of 1890, the UK swapped Heligoland with Germany for Zanzibar. More recently, in 1989 an unidentified Japanese company or individual tried to buy the Faroe Islands from the Danish government, in a deal involving restructuring of the islands' debts. The deal did not go through. In 1987, companies from South Africa, then still an international pariah because of the apartheid system, attempted to gain control of a freeport in the Isle of Man which would have made them very significant economically and thus politically in the island. In 1993 there were newspaper reports emanating from the Italian *Corriere della Serra* that the Mafia had bought up 60 per cent of land on the island of Aruba, a self-governing part of the Kingdom of the Netherlands off Venezuela. The Mafia family, it was reported, were also financing the election campaign of the ruling party (*Independent*, 5 March 1993). In the modern world it may even be companies that cause grief to powerless islands. In the late 1990s, there was considerable dispute between the USA and Europe over bananas, with the US-dominated World Trade Organisation forbidding the European Union's protection of banana imports from West Indian islands. The reason for the decision, it was widely thought, being pressure imposed on the WTO and the American government by major American-based multinational fruit companies with extensive interests in the large scale plantations of central and south America. 'The stench of rotten bananas' was one cynical headline (*Independent*, 14 September 1997) and the dispute much exercised the European parliament in early 1999 (see *EP News*, March 1999, p. iv). Traditionally more than half of the exports of St Lucia, St Vincent and Dominica come from bananas. An ex-Prime Minister of St Lucia said, 'bananas are to us what cars are to Detroit' (Ferguson, 1998). In an echo of the island as chess piece concept, the St Lucia Minister of Commerce, Industry and Consumer Affairs said 'Globally, we're just a lonely pawn on a gigantic chess board, surrounded by kings, queens and rooks who are waiting their moment to pounce' (*Time*, 26 July 1999).

Small islands' scale is such that they can participate in only few economic activities that involve off-island transfers of goods. These may well be at a level at which the island contribution is insignificant and, therefore, the island has no say in the operation of the economic activity and is the subject to decisions made elsewhere for the convenience of other actors. Thus an international hotel chain may decide to disinvest in an island for the sake of the company, and the island's loss of economic well being is simply an unfortunate, but disregardable side effect. This is the operation of globalisation, of course, and places through-out the world, on continents and islands both are subject to the whims of the transnational companies. But the familiar scale and resource limitations of islands make them especially vulnerable to such foreign domination. Mauritius would be an example. This island has benefited from the new international division of labour, which saw transnational companies seek cheap labour locations for manufacturing operations. As a result, Mauritius has become relatively prosperous. The authorities would now like to relax the low wage policy that, along with the decent education system and a multi-lingual workforce, had helped to attract foreign companies to the island. This has not proved easy to accomplish; foreign companies have intimated that if their labour costs increase they will simply move on to other locations where labour is cheap. Mauritius is thus subject to foreign domination of one of its principal economic activities, a domination that sees Mauritians' rewards for their labour held down.

In those cases where an island does not have a diversified economy, but concentrates on one export product, its dependence on outsiders can be near total. The classic case here is the decision of the British post office in the mid-1960s to switch from string made in St Helena to twine and rubber bands to bundle its letters. St Helena's productive economy was focused on the growing of flax as a raw material to make the string and the economy collapsed when the contracts were not renewed (see Chapter 10).

When islands get a measure of decision making, even if not necessarily full independence, it is interesting to see their people try to overturn some of the decisions imposed by outsiders against what the locals consider to be their better interests. Greenland had been taken by Denmark into what was to become the European Union in 1972. In 1979, the process of *hjemmestyre*, or home rule, granted Greenlanders more say in their own affairs and the island in 1985 disengaged from the nascent EU to be able to protect their fish stocks from having to be shared.

Islands dependent upon foreign aid might discover that the aid is linked to foreign domination. By no means all aid is given for purely altruistic reasons. The aid givers can seek political support, in the United Nations, say. They can link aid to trade, tying in islands to the purchase of goods from the donor country.

Islands, bases and the Cold War

During the Cold War, islands were particularly vulnerable to interference from America and the then USSR (or satellites), often ostensibly for trade and/or aid reasons. One coconut plantation owner on Vanuatu opined that:

> if the old colonial countries want to keep their influence here, they won't let the coconut economy die. Vanuatu is a woman who will sleep with anyone. And the customers are queuing up.

> (Evans, 1992, p. 18)

The Soviet Union's arrangement to operate a fishing fleet in I-Kiribati waters in August 1985 in return for an annual fee of £1.2 m was a flimsy veil for political expansionism. Earlier in 1985 the USSR had failed to get a similar deal with the Solomon Islands and it also negotiated with Vanuatu, Fiji and Tonga, seeking shore base facilities as well as fishing rights. Under American, also Australian and New Zealand, pressure, Kiribati did not renew the deal with the Soviet Union in 1986 and no other country signed an agreement with Moscow. Following a week of negotiations in Nuku'alofa, the capital of Tonga, in October 1986, the Americans, in recompense, agreed a five-year aid and development package worth $12 m per annum for the region, which included an annual £2 m fee from the US tuna industry. The American Tuna Boat Association, backed by the US government, had not recognised the 200 mile (360 km) Exclusive Economic Zones declared by the island states under the Law of the Sea treaty in the 1970s.

In the Indian Ocean there were strong rumours that the Soviet Union had established a military presence in the Seychelles. However, in 1984 President Albert René denied this, claiming that the Soviet-made missiles were part of the Seychelle's own defences and that North Koreans there were training the Seychellois army. The story resurfaced in 1987 when there were further denials. Certainly President René never granted the Soviet Union the deep-water naval port they wanted. Meanwhile, the US maintained a tracking station on Mahé Island, for which in the early 1990s they were paying £2.4 m per annum. Earlier, in 1981, the South Africans had attempted to establish a friendly regime in the Seychelles and mercenaries had been sent to Mahé. There have also been allegations of Mafia involvement in these islands (*Independent*, 9 June 1992).

The Americans, too, have used – and abused – islands, American territory or not, for their own geopolitical reasons. Many islands throughout the world have or have had US military bases. Many are not on American soil. There has been a US base on Cuba, at Guantanamo Bay, since well before the Castro era and it remained in operation throughout the Cold War, the Cuban missile crisis and America's decades long boycott of that island and its communist government. There are, or were until fairly recently, American bases on Bermuda; in the Turks and Caicos Islands; on Okinawa; in the Philippines; in Diego Garcia, which is part of British Indian Ocean territory; in the Marshall Islands; in Greenland; on Ascension Island. None of these islands are now American; most never were, but their authorities have been unable or unwilling to resist the Americans' use of their islands for their strategic purposes. The most controversial case relates to Diego Garcia in British Indian Ocean Territory. Here, at the Americans' request, between 1965 and 1973 the British forcibly removed the population of 1151, in some cases descendants of people who had been brought to the islands to work on the coconut plantation in the nineteenth century. In the 1980s the Americans spent about $500 m dredging the lagoon and lengthening the runway and Diego Garcia is now one of the Americans' most potent assets. Visits by civilians, including Britons, to this British territory are not normally permitted. Royal Naval Party 1002, comprising 43 British service personnel, is responsible for the laws and administration of the island, with the authority to arrest malefactors from within the 3500 Americans. In 2000 it emerged that the Americans may have offered the British an $11 m price reduction on Polaris missiles as part of the deal to gain access to Diego Garcia. In the same year some of the expelled people began a court case in Britain claiming they were unlawfully removed; the British government's position is that they were contract labourers. The High Court found for the islanders.

Some American bases on islands have closed: Bermuda, Kulusuk Island off eastern Greenland (Figure 7.2), the Philippines, and the Turks and Caicos Islands, for example. The Philippines bases, at Subic Bay and the Clark airforce base, were the largest employers in the country after the government. Many Filipinos wanted them to stay, but in the end sovereignty and national pride won the day and the Americans were obliged to leave Subic Bay by the Philippines government in October 1992. The previous year Clark had been rendered unusable by ash deposits from the eruption of Mount Pinatubo and the Americans had given it up. The Americans had prepared for the eventuality of having to leave the Philippines and had other facilities in the Pacific. In the 1980s, they were trying to make more use of their Trust Territory of Palau in the western Pacific, but a ruling by the island's government that Palau was a nuclear free zone in 1979 was causing difficulties. There has been considerable criticism both on the islands and internationally of the Americans' relationship with Palau (see, for example, the *Independent*, 6 January 1991).

The Americans are under pressure to leave Okinawa: 'Yankee go home' was one predictable headline (Desmond, 1995a). Part of the problem is that the Americans are simply too dominant in post-Cold War Okinawa, a territory they had held on to from after World War II until 1972. They control 20 per cent of the island's territory in 12 bases, including half the arable land. The US had, in 1995, 28,000 troops on Okinawa, out of a total of 45,000 in Japan, whilst there were only 37,000 in Korea. Further, there were problems over crimes committed by Americans and their dependants. Matters were brought to a head when a 12-year-old girl was raped by three American servicemen in 1995, after which 85,000 people, about 8 per cent of the population, protested the American presence in a mass demonstration in Naha. Polls showed that 80 per cent of Okinawans wanted them to leave. After the rape, the Okinawa Governor, Mashahide Ota, made the withdrawal of American troops a political priority and began to cause problems with the renewal of leases for land used by the bases (*South China Morning Post*, 15 May

Figure 7.2 Abandoned US equipment after closure of base, Kulusuk Island, Greenland, 1999

1996). The Tokyo government ordered him to co-operate. A non-binding referendum was held in 1996 ('Yankees, get lost' was the headline this time (Kunii, 1996)) and 89 per cent of votes were for the Americans to leave. After this the Japanese Prime Minister admitted that Tokyo had ignored Okinawa's plight, granted $47 m in aid and vowed to force the Americans to close down or move five of its installations (Gibney, 1996). By 2000 only one base, in Nago, had had its facilities relocated. By no coincidence that was the town where G8, the world's leading economic powers, met in July 2000 at a cost to the Japanese exchequer of £500 m. Mashahide Ota, no longer governor, spoke angrily of an attempt 'to buy out souls with money rather than treat their feelings and aspirations with respect' (*Independent*, 21 July 2000). Two other unsavoury incidents involving American personnel earlier in July had again rekindled local antipathy, and 25,000 Okinawans took the opportunity of the publicity brought by G8 to protest once more, by forming a human chain around the biggest of the US bases.

Presumably to avoid such problems and in the likelihood that there will never again be a compliant ally who will depopulate a strategic island for them, the Americans have kept some islands under their direct control. Guam, for example, is separately administered from all other islands in the central Pacific region and, unlike the others, will not move towards independence, but will presumably remain an unincorporated territory of the US as long as the Americans have need for their military facilities there. There is an important but controversial military area in Puerto Rico at the Roosevelt Roads base on the main island and on Vieques Island on the other side of the narrow strait. There were protests against the base in 1999 and people have died in the struggles here. However, despite such tragedies, a non-independent Puerto Rico presumably cannot force the US to leave these facilities. There are also bases on Hawaii. Hawaii has been an American state since 1959, after being seized by force in 1893 by US marines who overthrew the local ruler, Queen Lili'uokalani. Changing Hawaii's status from territory to state was controversial in the mid-1950s, partly because of fears that Hawaii would be open to communism. Senator J. Strom Thurmond, at the time of writing still in the Senate well into his 90s, took the argument a stage further, as explained in this passage by Gavan Daws:

> The clash of values that concerned him was one of much more ancient origin, expressed in culture, but rooted in biology. Hawaii could never be incorporated as a truly American state. The national body politic would reject it like some inassimilable alien substance. 'There are many shades and mixtures of heritages in the world, but there are only two extremes . . . our society may well be said to be for the present, at least, the exemplification of the maximum development of the western civilization, culture and heritage. At the opposite extreme exists the Eastern heritage, different in every essential – not necessarily inferior but different as regards the very thought processes within the individuals who comprise the resultant society'.
>
> (Daws, 1968, p. 388)

Thurmond went on to quote Kipling's 'East is East and West is West and never the twain will meet'. Thus Hawaii, a Pacific island chain, was to Thurmond too foreign to be allowed in to the USA. However, such views were of less significance than the realisation that America 'was a world power, with especially heavy commitments in the Pacific hemisphere, and Hawaii was an indispensable forward base' (Daws, 1968, p. 386). Unlike the situation on Okinawa or the Philippines, there is no governmental pressure on the Americans to reduce their presence in the Hawaiian archipelago. There are some Hawaiian people who would wish for more independence, and there are movements such as Ka Lahui Hawaii which have this agenda, but native Hawaiians make up only about 13 per cent of the total population.

There are other stories of island bases that, if space permitted, could have been told; the British on Malta and Cyprus, for example. The Malta story gives another twist to island powerlessness for it was against the wishes of many islanders that the British from 1967 cut back the bases on Malta, finally withdrawing in 1979. The Maltese economy was troubled by the loss of jobs and political relations between the UK and Malta were severely strained (Blouet, 1989).

Islands and their political status

Politically, islands now range across the spectrum from total dependence to complete independence. The fact that all islands were once dependent has been dealt with; this section looks now at the current situation.

Restricted local autonomy

Small islands, particularly those on the continental fringe, and especially if they are not near any other islands, tend just to be assumed into the nearest mainland local government unit. Northern Ireland has local government districts as the lowest tier of political authority and here the one offshore inhabited island, Rathlin, is just part of Moyle District Council. In the Republic of Ireland the fringe of islands are just part of the relevant County Councils with few, if any, political powers reserved to the island, beyond community activities. Some of the counties administer a number of islands and have set up special committees to oversee island affairs; Mayo and Cork are examples, but such arrangements did not go far enough to satisfy many Irish islanders. Political powerlessness is just one of a common range of challenges faced by islanders throughout Ireland, as was realised by them in the mid-1980s, when, led by some of the island co-operatives, people from the different islands began to get together. A pressure group, the *Comhdháil na nOileán* (Council of the Islands) was set up (Royle, 1986). One of its demands was that the political structures of Ireland be adjusted to grant the islands some form of collective recognition. That original body faded but the idea remained, and was taken up by a successor group *Comhdháil na nOileán na hEirann* (Council of the Islands of Ireland) which was part of the successful campaign to achieve the establishment of a government committee to oversee Irish island affairs. The committee's mission statement was:

> to support island communities in their social, economic and cultural development, to preserve and enhance their unique cultural and linguistic heritage, and enable the islanders to secure access to adequate levels of public services so as to facilitate full and active participation in the overall economic and social life of the nation.
>
> (Department of the Taoiseach, 1996, frontispiece)

Now the process has gone one stage further and there is a Minister of Arts, Heritage, Gaeltacht [Irish language] and the Islands. Meanwhile the *Comhdháil na nOileán na hEirann* continues its work, and in 2000 brought 50 Europeans to Inishere to begin work on the establishment of a pan-European island pressure group.

  Elsewhere, there are islands with more political power either because they are of larger size or because they form convenient groups. Still in the British Isles, Anglesey (Ynys Môn) and the Isle of Wight are counties; Shetland, Orkney and the Western Isles are Island Regions of Scotland. The Isle of Man and the Channel Islands (separately as the Bailiwicks of Jersey and Guernsey (which includes smaller islands such as Alderney)) are British, but are not part of the United Kingdom and have much local responsibility under the British crown. In Washington State in the USA, there is an insular county, the San Juan Islands. The Åland Islands are part of Finland but are Swedish-speaking and self-governing. Other European island groups, which have become autonomous units, include the Balearic Islands and the Canary Islands of Spain, Madeira and the Azores of Portugal, Sicily and Sardinia of Italy and Corsica of France.

At a larger scale, two Canadian Provinces (Prince Edward Island and Newfoundland) and one Australian State (Tasmania) are insular. Newfoundland Island has even the unusual distinction of administering part of the Canadian mainland, in Labrador. Newfoundland may not be well developed by Canadian standards, but Labrador, with the exception of some mining settlements and coastal communities, many of which are inhabited by First Nations groups, is barely developed at all. Here is the usual power relationship: the relatively strong overseeing the relatively weak; unusually in this case it is the relatively strong island which has the dominant role. Other large Canadian islands are not provinces. These include many in the north, too remote and sparsely populated for provincial status, but in the east there is an island of equivalent size and population to Prince Edward Island but which is just part of Nova Scotia. Thus is Cape Breton. It had a separate history from the mainland part of Nova Scotia in that its ownership transfers between Britain and France in the long-standing competition for territory and possessions in Canada between these two nations were different. It finally finished up as a British colony under the Provisions of the Treaty of Paris in 1763. However, Cape Breton was then overwhelmed by migration from Scotland, displaced by the Scottish Highland Clearances, and had to be joined to Nova Scotia in 1820 to scale up the resources available to deal with the migrations. The Canadian and Australian experiences show that islands become provinces/states of federal countries only when historical circumstances are favourable to this. Newfoundland, Prince Edward Island and Tasmania are all relatively poor within their nation states and are the smallest provinces/states in population terms.

Moving onto the national scale there are only two continental countries that have their capital cities on isolated offshore islands, Denmark and Equatorial Guinea. As explained in Chapter 3 both these cases of the political power being sited on the island can be explained in terms of their historical development. In all other countries with continental and insular territories, the capital is on the mainland, including those with substantial insular realms such as Malaysia and Greece.

Non-independent islands tend to have a restricted political status, with their mainlands exercising considerable power over them. Sometimes this causes resentment and there are a number of calls for independence from islands wanting free of the mainland yoke. The Canary Islands and the Balearics have political pressure groups seeking independence from Spain (see, for example, Hornblower, 1991); on Corsica, independence pressures have degenerated into violence through the activities of FLNA and other groups against French rule. There was also a recent suggestion that if the UK entered the European Monetary Union, the Isle of Man would try to secede on the grounds that tax harmonisation would jeopardise its financial services industry.

Multi-island nations also had and have secessionist pressures. The West Indian Federation was never successful; at a smaller scale in the West Indies, the citizens of Nevis were polled in 1998 on whether they wished to secede from St Kitts and Nevis. The majority was for secession but not at the 60 per cent level needed for change to take place. In the Pacific, the people of Bougainville, ethnically Solomon Islanders, have never been content at being part of the independent state of Papua New Guinea, just one of the problems that have troubled this island, whose environment has been blighted by copper mining (Connell, 1997). Violence against the mine led to its closure; violence against the national government led to the island being blockaded to the detriment of its economy (Fathers, 1992). Anjouan islanders wished to secede and declared their independence from the other Comoros Islands in 1997, but troops from the principle island in the chain, Grande Comore, invaded. In 1974, Mayotte, the fourth island in the chain had refused to move towards independence with the other islands and remains a territorial collectivity of France. At a different political level, Staten Island in 1993 started the process of trying to leave New York City in order to have more local power. Also Key Biscayne, an affluent island on the Florida Keys, is trying to leave Miami's Dade County in order to be able to set up its own planning regulations.

This section has shown that dependency is not always a stable situation regarding the political structures of islands. In other cases instability has resulted in the division of islands as will now be discussed.

Divided islands

There is a geographical wholeness attached to an island, and normally this sees it treated as a unit in political terms. Be it with its own status or as a dependent part of a larger political unit, islands are normally not divided politically. The Canary Islands might form two separate Autonomous Regions of Spain, but these two groups are each made up of whole islands. However, there are a few exceptions with regard to both local and national government. For example, the new Canadian Territory of Nunavut has boundaries that cut through Victoria and Melville Islands in the Arctic, the other parts of which remain in the Northwest Territories. However, these islands are barely populated and cannot be said to function as island economies or societies. More inexplicable was the historic division of one western Scottish landmass into the Isles of Lewis and Harris, which until the 1970s were in different administrative areas – Lewis was part of Ross and Cromarty whilst Harris was in Inverness-shire. In the reorganisation of Scottish local government in the 1970s this political division was ended when Lewis and Harris joined the other Outer Hebrides in forming the Western Isles Island Council (*Comhairle nan Eilean*).

This unification paralleled the situation on the larger island to the east of Lewis and Harris, i.e. Great Britain, which must be briefly mentioned here for the sake of completeness, though hardly a small island. Great Britain also used to be divided but Wales was absorbed politically into England, though retaining many cultural differences, from 1277; and Scotland joined the Union when its King, James VI, became also King James I of England in 1603. Scotland retains more differences from England than Wales, with its own legal and educational systems. Both Scotland and Wales tend to compete in sports, for example rugby and football, as national teams, separately from England. However, such things, however significant culturally and in the popular imagination, do not amount to political independence – and at cricket, Welsh and Scots players have, on occasion, been selected to play for the UK's only Test playing unit, England. In 1999 more regional autonomy came to Scotland and Wales with the setting up of a Scottish Parliament and a Welsh Assembly. Although in both cases nationalist parties won many seats at the elections, they did not gain power and the new structures do not amount to independence. The UK parliament at Westminster retains overarching authority and the island of Great Britain remains unified in political terms at the nation state level. (The other part of the UK, Northern Ireland, will be considered below.)

In fact, only very rarely are islands divided between nation states but these unusual cases are part of the island political story and must be dealt with here. The cases are Borneo, Cyprus (*de facto*), Hispaniola, Ireland, New Guinea, Saint-Martin/Sint-Maarten, Tierra del Fuego and Timor (Figure 7.3). These divisions are almost all the results of colonialism. Borneo is huge, at 743,107 km^2 and was not subject to any overall political control in pre-colonial times. It then got caught up in the contestation between different colonial spheres of influence; the south being Dutch controlled, much of the north at least nominally in the hands of the British. These areas have moved to independence as parts of Indonesia and Malaysia respectively, with, uniquely, a third nation forming part of the island in the form of the oil rich territory of Brunei, which was a separate British-protected sultanate until its independence in 1984. New Guinea, similarly, was in myriad tribal hands until colonialism which saw it, too, at one time in three political units with the Dutch holding Irian Jaya in the west, the British, Papua in the south-east and the Germans, New Guinea in the north-east. The Germans lost their empire in World War I and New Guinea was administered by Australia from 1914–21 and then by the British until 1945 after which it was joined with Papua and Papua New Guinea became independent in 1975. Irian Jaya is now part of Indonesia, but was retained by the Dutch until 1963, the rest of Indonesia having become independent in 1949.

Indonesia has also part of what is a third divided island – Timor. In the east of this island was a Portuguese colony, East Timor. However, in 1974 when the Portuguese Empire was fast disintegrating into anarchy and civil wars as that country underwent revolution at home and took its eye off the ball overseas, East Timor was invaded by Indonesia. Indonesian rule here has been very controversial, many have died and there is little meeting of minds between the Catholic East Timorese and the Moslem Indonesians (see, for example, Barbedo de Magalhães, 1992). In 1999, after much unrest, there was a UN supervised referendum in East Timor in which voters expressed a desire for independence. There was then considerable violence, which led to foreign intervention, led by Australia, upon which the Indonesian troops withdrew. East

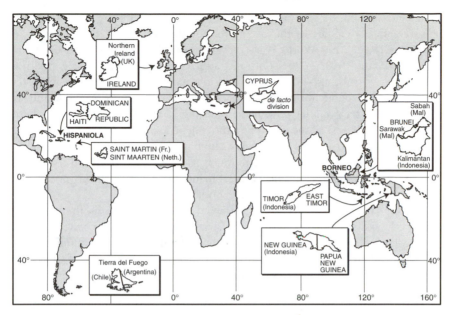

Figure 7.3 Divided islands

Timor now will presumably move to full independence, forming, once again, a divided island.

The activities of European colonialists also saw the division of Hispaniola between formerly French Haiti in the west and what has become the Dominican Republic in the east, once a Spanish possession. This process was mirrored at a smaller scale in the tiny West Indian island of Saint-Martin/Sint-Maarten where Dutch and French interests met and, unusually, the island's territory was shared. It is still colonial with the north being part of the French Départment d'Outre Mer of Guadeloupe and the south being part of the Netherlands Antilles. Colonialism also saw the barely inhabited island of Tierra del Fuego at the tip of South America split between Argentina and Chile. Like many of the other divided islands, Tierra del Fuego had no overall authority to resist the colonialists. Its international boundary is for the most part a classic colonial one; a straight line running north–south, rather like that which divides New Guinea.

Now we come to Ireland. To detail the long saga of the Irish partition is outside the scope of this volume, which anyway has a focus on smaller islands. However, for the sake of completeness, again, something must be said. The story, as always, is of relative powerlessness. The most significant event regarding the island's present division was the Plantation of Ulster in the seventeenth century. This saw the island's northeast corner assume a majority population of Protestants originating from Great Britain, mainly Scotland, in contrast to other parts of the island where the population was predominantly Gaelic and Roman Catholic. Ireland periodically rebelled against its domination, in 1641 and 1798 especially, but from 1801 was joined even more firmly to Britain when the parliaments united. There was periodic home rule strife during the nineteenth century and at the start of the twentieth, culminating in the Easter Rising in Dublin in 1916. After World War I it was impossible for Britain to deny the majority of the Irish – the Catholic and Gaelic section – their political independence whilst also politically impossible to force the Protestants in Ulster to join in an independent united island. Partition then became inevitable and six of the nine Ulster counties remained inside the United Kingdom as Northern Ireland whilst the rest of the island became independent. Partition did not bring peace or stability to Ireland and the last three decades of the twentieth century saw the 'Troubles', associated with much violence and death in Northern Ireland and elsewhere as a result. Political manoeuvrings over the governance of Ireland are being actively pursued as this book is written.

Cyprus' partition is internationally unrecognised. Its cause is the common one of ethnic division. This island, in its strategic location at the eastern end of the Mediterranean, has long been part of overlapping spheres of influence from Greece and Turkey. It ended up with a population largely made up of people with ultimate ethnic origins from those two countries, the majority being Greek Cypriots. Upon independence in 1960, from Britain – the last in a long series of colonial rulers – Cyprus had a power sharing executive, but the minority Turks felt discriminated against and there was a good deal of strife. In 1974 a coup put a hard-liner in charge of Cyprus' government and Turkey invaded, ostensibly to protect its ethnic fellows. Cyprus' guarantor powers did not intervene, and by the time the UN were able to negotiate a cease-fire, the Turks had taken about one-third of the island. There was a population exchange and since that time Turkish Cypriots and mainland Turks have lived to the north of the line and Greek Cypriots to the south. The line has become, *de facto*, an international border, although only Turkey recognises the legitimacy of the Turkish Republic of Northern Cyprus, which administers the north (Figure 7.4). There are periodic attempts to reconcile the two sides on the island but nothing has yet come to fruition.

One other island, still theoretically united as a political entity, is in many ways divided in practical terms. This is Sri Lanka, a once colonial island that like so many others ended up with an ethnically diverse population. Here the major division is between Sinhalese (c. 11.8 m) and Tamils (c. 2 m), with the former controlling most of the island's governance post-independence. The Tamils have been pressing for an independent homeland, Tamil Eelam, in the north and northeast of the island where they are concentrated. Violence broke out in the 1970s and many thousands have died. Sri Lanka was subject to outside intervention in the conflict, thus from 1987 to 1990 up to 70,000 Indian troops were stationed there as a rather unsuccessful peacekeeping force, themselves losing many men.

What becomes clear from this brief description of the world's divided islands is that each of them became divided after interference from outside, be this colonialism, migration, or invasion – sometimes all three: island powerlessness again.

Figure 7.4 The border of the Turkish Republic of Northern Cyprus, Nicosia, Cyprus, 1993

Contested islands

Many of the divided islands mentioned above became subject to partition through contestation, often expressed violently. Often the resultant boundaries have become universally accepted and there is not pressure to change. Nobody, to the author's knowledge, struggles for the unification of Borneo, or of Tierra del Fuego, Hispaniola or New Guinea. There are people in Irian Jaya who want freedom from Indonesia, but the unification of the whole island of New Guinea is not on the agenda. However, elsewhere there has not been quiescence since partition and people have died in Cyprus and Ireland in struggles for political unification. This brings up the point that islands, even though not always attractive places from an economic viewpoint, can be and are subject still to contestation, to conflicting claims. Many of these disputes are familiar and of long standing, still causing trouble today: Cyprus, Ireland, Timor, etc., and they have already been introduced above. Others have yet to be mentioned and bring to light a new political category, the contested island.

Here we are not dealing with the islands where past contestation has left a legacy of what internationally is considered to be domestic political strife. Corsica is not contested between nations; the island is universally recognised as part of France however much some Corsicans resent this. Instead we will look at islands whose very ownership is disputed internationally in the way that Spain still contests Britain's ownership of Gibraltar. There are a number of examples, some between friendly countries. The International Boundaries Research Unit have publications detailing disputes between Indonesia and Malaysia over Pulau Sipadan and Pulau Ligitan (Haller-Trost, 1995); between six nations over the Spratly Islands (Dzurek, 1996), as well as a general text entitled *Island disputes and the laws of the sea* (Smith and Thomas, 1998). Further, the Americans and Canadians both claim Machias Seal Island in New Brunswick (or is it in Maine?); maritime boundaries in the Gulf of Maine have long been disputed between these allies (Ricketts, 1986; Burnett, 1990). The French do not recognise Les Minquiers as part of the British Channel Islands; Wake Island is part of the Marshall Islands to the Marshallese, but not to the Americans who have a base there. The Marshallese are hardly likely to press their case; the Americans are the nation with whom they remain 'in free association' after their independence and who they very much need to provide aid.

Such disputes are in many cases not of great import but there are some really serious problems, too. Greece and Turkey, despite being NATO partners, could not be described as friendly, although relations improved in the very late twentieth century when the two countries helped each other after both were affected by earthquakes. One cause of dispute between the two is the contested ownership of Imia (to the Greeks) or Kardak (to the Turks). There are about 1000 rocks and small islands in this area, all uninhabited. Their ownership had been left unclear after the cession to Greece of the Dodencanese archipelago in the southeast Aegean from Italy in 1947. So say the Turks. The Greek view is that the islands became Greek in 1947. Contestation for these rocks saw the mobilisation of naval forces as recently as January 1996. One version of events has a Turkish ship, the *Figen Akat*, becoming grounded there on Christmas Day 1995 and the Greeks sending in troops to assert ownership, with the Turks responding in kind (*Turkish Daily Press*, 19 July 1996). Alternatively, a Greek Orthodox priest went to the islands and raised a Greek flag, which was taken down by a Turkish media group who raised a Turkish flag. Next day the Greek military arrived (*Independent*, 1 February 1996). What is not disputed is that twenty warships were involved and a Greek helicopter was lost to a crash during the incident and that the situation almost led to war. However the US political dispute negotiator, Richard Holbrooke, cooled the situation.

The dispute between the UK and Argentina over the ownership of the Falklands is widely known, has been discussed earlier in this chapter and has, perhaps, some way to run yet. In 1998 Klaus Dodds wrote about the dispute still being 'unfinished business', although to many of the islanders any business with Argentina is over. The point is, though, the Argentinian claim remains and has to be countered. Dodds noted that in the 1990s the UK was spending around £67 m per annum on defence (1998, p. 623). In July 1999 it was announced that better arrangements had been made with Argentina over fisheries, that Argentine citizens will be able to visit the Falkland Islands on their own passports, and that from October 1999 flights to the islands from Chile would stop in Argentina. This announcement saw street protests in Stanley countered by the release of a press statement by the Falkland Islands government explaining the benefits of what had been agreed (Falkland Islands Legislative Council, 1999). Unfinished business indeed.

Another dispute that has seen military engagement and death relates to the Spratly Islands, 230 shoals, spits and reefs scattered over 180,000 km^2 in the South China Sea. These islands are basically reefs, only some of which remain dry at high tide. Some of them have been utilised as temporary refuges by fishermen but they are not permanently habitable. Nonetheless, no fewer than six nations claim all or some of the islands: Brunei, Malaysia and the Philippines claim part; China, Taiwan and Vietnam all of them (Figure 7.5). Their disputed ownership is a major source of difficulty to the working of the regional political grouping ASEAN, as seen at the 1992 meeting in Manila and that of 1994 in Bangkok. The islands were, in modern times, first under the influence of France, operating from Vietnam, and were taken over by Japan during World War II. Japan rescinded its claim at the San Francisco Peace Conference in 1951. Five of the claimants (all but Brunei) have occupied some of the islands with military forces, sometimes by building structures on stilts to gain living space above water level. Other nations have reacted violently to such ventures and there have been battles. In 1988 the Chinese sank three Vietnamese ships when they took six reefs from Vietnam. Seventy-seven men died. In 1994 some Filipino fishermen were seized by Chinese warships on the Spratlys. In 1997 Filipino fishermen and congressmen hoisted a flag over a shoal and China protested at this infringement of sovereignty. In 1998 China began to build two concrete 'fort-like' structures on the aptly named, Spratly sub-surface

Mischief Reef (Meijijiao to the Chinese) to the discomfiture of the Philippines, within whose 200 mile exclusive economic zone Mischief Reef (Panganiban to the Filipinos) lies. Defense Secretary, Orlando Mercado, fears 'creeping invasion' by China, with the building of a Great Sea Wall. China claims possession of the Spratlys, given that Ming dynasty admiral, Cheng Ho, sailed several times in the region in the early fifteenth century, and said that the structures were being built by fishermen (McCarthy, 1999). China had, anyway, passed a law in 1992 claiming the Diaoyutai, Spratly and Paracel Islands, in response to a joint Vietnamese/Malaysian proposal to develop some of the Spratlys. There were serious disputes, too, over the other groups.

China in 1996 warned Japan of 'serious damage' if any more Japanese right wingers set foot on the uninhabited, disputed Diaoyutai Islands (Senkaku Islands to the Japanese) on which the Japanese Youth Federation had built a small beacon. The Americans administered the islands after World War II, but had ceded them to Japan in 1972, but both China and Taiwan disputed this. The islands have also been claimed by Taiwan (as the Tiaoyutai Islands). The Taiwanese threatened to let fishermen tear down the beacon; in 1990 Japanese patrol ships had stopped two Taiwanese boats from landing. In the event in 1996, Hong Kong Chinese went out to the islands to try to dismantle it and one of the party drowned. Japanese goods were boycotted by some in Hong Kong as a result of the dispute.

China seized the Paracels in 1974, islands also claimed by Taiwan and Vietnam. In 1994 Vietnam passed a resolution reaffirming its ownership over the Spratly (Truong Sa) and Paracel (Hoang Sa) groups. Also in the region, Japan disputes with South Korea over sovereignty of the Takashima (to the Japanese) Islands.

Elsewhere in recent years, Honduras and El Salvador, two countries that famously went briefly to war in 1969 over a football match, dispute ownership of Meanguera Island. Qatar and Bahrain had a military flare up over Fasht al-Dibal reef in 1986, during which 29 foreigners were kidnapped. The dispute was mediated by the Saudi ruler, King Fahd. Qatar and Bahrain also dispute Huwar Island. Also in the Gulf, Iran disputes with Sharjah Abu Musa and with another emirate Ras al-Khaimah over Greater and Lesser Tunb Islands. In 1995 Eritrea and Yemen came to blows over Greater and Lesser Hanish Islands and both countries suffered casualties.

Why on earth should people die over such scraps of territory? Because today islands present their political owners with the rights to claim territorial waters; such waters give the nations access to the marine resources: fish, perhaps minerals including things like manganese nodules found on some seabeds, and in some areas the possibility of oil. The Spratlys are already subject to oil exploration. They, plus Diaoyu/Senkaku/Tiaoyu and the Paracel Islands, remain potential flash points for the twenty-first century.

Of greater international significance because of the power of the nations involved are disputes over islands between Russia and the USA, or Russia and Japan. Russia now rather regrets its sale of Alaska to the USA in 1867 for two

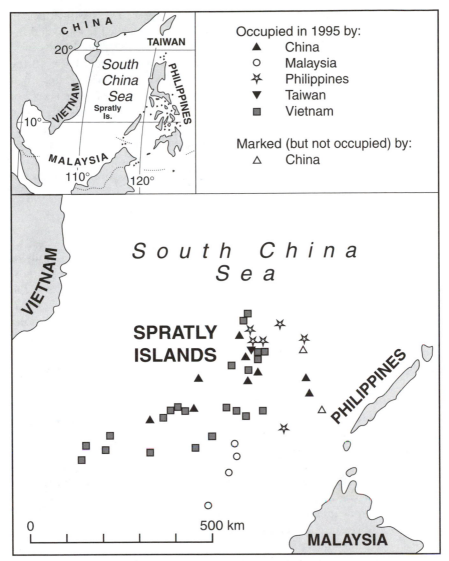

Figure 7.5 Contestation in the Spratly Islands, South China Sea

cents an acre and claims back another appropriately named island, Wrangel Island. The Russo-Japanese dispute is over seven of the Kurile Islands, Etorofu, Kunashiri, Shikotan and the four-island Habomai group. The Kuriles stretch from the Kamchatka Peninsula to Hokkaido, and Russia took these, the islands closest to Japan, after Japan's surrender in 1945. The two countries had disputed ownership of the Kuriles since the eighteenth century when Russian fur trappers first encountered the islands' aboriginal Ainu people. An 1875 treaty had given Japan the Kuriles and Russia Sakhalin Island. Japan always disputed the legality of the 1945 seizure, and did not sign a peace treaty with Russia in consequence of this dispute, although in 1956 Russia offered Kunashiri and the Habomai Islands to Japan if it would. In 1992 again the same islands were offered to Japan, this time in return for a Japanese aid package. By 1948 the 17,000 resident Japanese islanders had been expelled, presumably against their will, by the Russians. In a 1991 referendum, 75 per cent of their successor islanders, 25,000 Russian civilians, had voted against the islands becoming Japanese, but their country's leaders were prepared to return the islands anyway if the deal was right for Mother Russia. One newspaper report noted that 'none of the islanders expected their interests to be uppermost in the minds of Japanese and Russian diplomats' (*Independent*, 2 July 1992): island powerlessness again.

Island 'colonies'

In 1998 Robert Aldrich and John Connell published a comprehensive study of dependent territories called *The last colonies*. The existence of this book almost makes redundant anything that might be said here about island colonies in the few hundred words that space permits. However, for the purposes of this volume it must be pointed out that one reasonably common form of insular political status is that of a 'colony'. The term 'colony' carries with it negative connotations of white, exploitative rule over unwilling and exploited subject peoples who are not white. That is not the situation in most of the world's remaining colonies today, many of which actually now benefit from their subject status. In fact the term colony is now rarely used, even 'dependency' is now not politically correct; instead 'overseas territories' is to be preferred. Thus, that is the term now for what were British Dependent Territories, previously colonies, according to a Government White Paper of 1999 (Foreign and Commonwealth Office, 1999).

Aldrich and Connell (1998) note that there is no definitive list of the last colonies. The US Department of State in 1998 identified 43 permanently inhabited dependent territories (cited from a website by Christopher, 1999, p. 327). Aldridge and Connell rather identify 54 places but their choice might be challenged: they include the Canaries which most people would regard as just an autonomous region (to be pedantic, two autonomous regions) of Spain, like the Balearic Islands which are not included. They do not include Svalbard (Spitzbergen) which has a dependent relationship with Norway, nor do they count Bouvet or Peter 1 Islands, uninhabited Norwegian dependencies in the Southern Ocean. The detail of what should be included is not important really; what does come from the last colonies, the book or the real world, is the fact that almost all places that are eligible to be considered for inclusion in such a category are islands. The great empires have gone, broken up into scores of nation states. What remains, with very few exceptions, are islands, too small, too weak, too defenceless to manage in the modern world without the guaranteed help that a dependent relationship with a metropolitan power grants them. Or, alternatively, perhaps, they are too useful to their metropolitan power for a change in their status to be countenanced. Two of the exceptions, mainland dependencies, were removed from the list in 1999 when the US handed over control of the Canal Zone to Panama and Macau reverted to Chinese rule, having been Portuguese since 1557 when it was leased from China.

For islands to remain as colonies does have drawbacks. One is a lack of self-regard, the 'nation yearning to be free' idea. Small islands engender a strong sense of identity, stronger in fact than many of the artificial and unstable so-called nation states bequeathed to, say, Africa by colonialism. There is little sense of national identity in the Sudan; the Hutus and the Tutsis in Rwanda hardly gelled into 'one heart, one people' to quote from a hoarding the author once saw extolling nationhood in Jamaica. Yet Falklanders, Kelpers, know who they are (and what they are not) despite not being 'free' in the sense that the tribally-, not nationally-identifying citizens of Nigeria are free.

There may be more graduates from *The last colonies* list, although the last island nation to graduate to independence was Palau in 1994. France has been a reluctant decoloniser and retains territories that would certainly prove to be as successful independent nations as many. There have been political movements in many of their islands to press for independence, sometimes accompanied by violence. One person died in a bombing in Guadeloupe in 1985, when at least two pro-independence movements were active in the island. The situation in New Caledonia in the mid-1980s was very serious (Connell, 1988). The aboriginal peoples, Melanesians known as Kanaks, are a minority within the islands at 43 per cent of the near 150,000 people. Most of them favour independence. The majority, mainly white settlers of past or recent French origin, wishes to remain French. The two sides have clashed violently. France held a referendum on future status in 1987, which favoured the status quo by a huge majority, 98.3 per cent being in favour. This hardly settled the issue, however, as the pro-independence grouping, the Kanak Socialist Liberation Front (FLNKS) had boycotted the vote (see chapter 8 of Aldrich and Connell, 1992). It was decided that there would be a long cooling off period before the issue was debated again (the Matignon Accord), but the separatists' leader, Jean-Marie Tjibou, was assassinated in 1989. Recently the debate was re-opened and in April 1998 Kanaks and settlers 'signed a deal with Paris which envisages a 15 to 20 year period during which powers would be gradually transferred to New Caledonia prior to a referendum on independence' (*Fiji Daily Post*, 2 May 1998). In French Polynesia, there is also an independence movement, *Tavini Huiraatira*, which became very active in 1995, partly to protest the French use of the islands for nuclear bomb tests. A Greenpeace spokesman noted that 'it's very clear that questions of independence and nuclear protests are linked' (*Independent*, 9 September 1995).

There are also pressures for independence from some other islands such as the Balearics and the Canaries in Spain. In Denmark, there is a nationalist party on Bornholm, *Bornholmus Fremtid*. In April 1999 there was an international conference on Microstates held in the Faroe Islands, where it was made clear that the Faroese were at least actively exploring the possibilities of seeking independence from Denmark. In 1999 it was announced that oil exploration was to begin in that portion of the Atlantic between Shetland and the Faroes. Finds on the Faroese side of the line would certainly help any progress towards independence, one problem being the considerable subvention the islands need from Denmark to maintain their standard of living. It does not take that much money in absolute terms to maintain 47,000 people, even if they are settled on small, remote islands.

Puerto Rico, Bermuda, Guadeloupe, Réunion, New Caledonia, Cayman Islands, Martinique, French Polynesia, the US Virgin Islands, American Samoa as well as the Faroes and, perhaps, Greenland: all these places would form microstates at least as viable as some of those island nations that are independent. The Marshall Islands, the Federated States of Micronesia, Kiribati, Nauru, St Vincent, St Lucia, especially Tuvalu, all nominally independent, have considerable problems to face in economic, social, resource and/or environmental terms. Perhaps that is why many of the remaining overseas territories have, in recent years, not opted to change their status; they are insulated from some of these pressures, if at the cost of their political freedom. This head versus heart argument was rehearsed in the run-up to the 1999 referendum on the political future of East Timor. There was an article in Canada's premier national newspaper by Marcus Gee advising the East Timorese to remain within Indonesia, although with a degree of self-government. Gee's reasoning was largely to do with economics:

> Independence ... is a leap in the dark. The independent nation of East Timor would be a flyspeck on the world map. Its only significant export is coffee. It has next to no industry. Its per capita economic output was $138 (US) in 1997, one of the lowest in the region. Independence would mean the end of Indonesian economic subsidies – only one-40th of the provincial budget is raised locally – so East Timor would have to throw itself on the mercy of the international community.... The only way to satisfy [the] yearnings [for independence] without producing rivers of blood is to try to accommodate them within the boundaries of existing nations – the solution we in Canada call federalism.
>
> (*Toronto Globe and Mail*, 25 August 1999)

Gee was answered not just by the vote, which went against his advice, but also by a letter from David Webster. This pointed out that in a globalised world the size of economic units was becoming of less importance and East Timor was more likely to flourish apart from the troubled Indonesian economy than 'shackled to it'. But most significant was the 'heart' argument; 'there was an inspiring story of human freedom to be told in East Timor' (*Toronto Globe and Mail*, 28 August 1999).

Often, though, it is the head that rules, though let it be acknowledged that few dependent territories had been through the trauma endured by the people of East Timor. In the French Antilles less than five per cent of voters opted for independence in the last referendum. In the Netherlands Antilles, the islands in 1993 or 1994 opted for the status quo by large majorities: 91 per cent in St Eustatius; 88 per cent in Saba; 86 per cent in Bonaire and 59 per cent in Sint-Maarten. Aruba, which left the Netherlands Antilles in 1986 to make progress towards independence, has since changed course. Bermuda (1995), Puerto Rico and the US Virgin Islands (1993) all had referenda on sovereignty issues recently and in each case their people voted to remain dependencies, 3 to 1 in Bermuda and 10 to 1 in the American territories. Christopher opined that the government of Bermuda (like that of Aruba), which had initiated the constitutional negotiations for independence, 'drew back when confronted with the full economic and political consequences of their actions' (1999, p. 328). If there is to be any change in Puerto Rico's status, it is likely to become the 51st US State rather than an independent country. Earlier, a similar referendum took place on the Cocos Keeling Islands in the Indian Ocean. These islands were granted as a private fiefdom to the Clunies-Ross family by Queen Victoria in 1886, Captain John Clunies-Ross having first occupied the islands in the 1820s. The 27 islands became an Australian territory in 1955 but were still owned by the family. Australia bought them out in 1978 for £3.5 m and amongst the changes made was the introduction of proper currency (the Australian dollar) for the first time. In April 1984 the few hundred islanders, descendants of indentured labourers brought in by the family to work the coconut plantations, were offered the choice of integration with Australia, free association or independence. They did not even opt for the halfway house of free association, but voted overwhelmingly in the UN supervised poll to become part of Australia.

Jerome McElroy has written on the benefits of island colonialism, noting that there are:

> substantial socio-economic benefits associated with political affiliation . . . free trade and export preferences . . . lucrative grants and social welfare assistance, ready access to off-island capital . . . in many cases the availability of off-island labour markets through migration, aid-financed infrastructure and communications, relatively high quality health and educational systems, natural disaster relief, the provision of costly external defense, and even protection against potential internal disturbance.
>
> (1998, p. 33; see also Royle, 1992)

Using comparative statistics from the Caribbean, McElroy went on to note that in matters such as GDP per capita, electricity consumption, motor vehicle and telephone ownership rates, dependent islands outperformed the independent island states. Similarly, population, social and educational measures favoured the non-independent islands. 'It is small wonder then that the process of decolonisation has stalled at the steps of the small-island Caribbean' and here and elsewhere in the small island realm he opined that we have entered the 'era of colonial permanence' (McElroy, 1998, p. 33; see also Connell, 1994). This ties in with the thoughts of Aldrich and Connell who conclude their long and thoughtful study of *The last colonies* by also assuming that the colonial era will probably never end. Globalisation has probably made independence less relevant anyway.

> People will protest against the iniquities of a perceived colonial presence. But in many territories, similar declarations will be more evidently rhetorical devices, designed to achieve less unbalanced links with the global economy, greater degrees of sovereignty, autonomy and cultural identity, but not political independence. Formal political independence now means little. The 'last colonies' have become the avatars of the post-modern future.
>
> (1998, p. 251)

Support for this view came from Christopher (1999) who noted that the debate on the economic and political advantages of independence for the remaining colonies 'had inclined towards retaining the status quo' (p. 328). In fact, it has been suggested that if forced to it, the Turks and Caicos Islands would see if they could become a territory of Canada, with which there were historic links because of the salt industry, rather than be independent. Further, there was a semi-serious suggestion in 1993 that the Pitcairn Islanders, dissatisfied with British rule, might like to try becoming French.

Islands as independent nations

The final possibility for an island's political status is to be independent. Few islands are independent by themselves. Most insular states are archipelagos, many with hundreds, even thousands of islands. In such circumstances a small island can have the same dependent relationship with the power centre of its insular state as an offshore continental island has with its mainland power base. Political independence in the case of many islands does not necessarily lead either to economic independence or much influence in world affairs. Small island states are often keen members of regional or other international groupings, trying to scale up their influence. Regarding economics, in island realms such as the South Pacific many islands are classically cases of MIRAB economies despite independence (see Chapter 8). Cameron investigated the Federated States of Micronesia to see 'is there a Pacific way to avoid a MIRAB society?' (1992, p. 150). His conclusion was that migration, under the terms of the compact with the US,

> will add the 'MIR' to the present 'AB', as labour moves to capital rather than capital moving to labour. Once more, independence will become little more than nominal, even if formally it may be real enough.
>
> (Cameron, 1992, p. 167)

Independent islands may also have inherited from their colonial pasts a divided society, leading to troublesome political divisions post-independence. Fiji is the classic example here, but there are many other examples, thus in early 2000

Figure 7.6 Cartoon marking 30 years of Jamaican independence, 1992

Source: *Jamaica Daily Gleaner*, 20 August 1992

there was much violence between Muslims and Christians in Malaku (the Spice Islands), Indonesia. Problems might also arise because of the possibility of a charismatic leader becoming too dominant in a small society, although the post-colonial history of Africa shows that this is not just an insular phenomenon. 'Papa Doc' Duvalier of Haiti was perhaps the most notable island example. Finally, small islands may struggle politically as their leaders face the serious issues of managing a small island economy. The cartoon from the *Jamaica Daily Gleaner* (Figure 7.6) exemplifies this.

Thus it seems that islands, no matter their political status, cannot escape from their basic geographical constraints, a theme that is taken up in the next chapter, when it is applied to island economies in the contemporary period.

References

Key readings

The two books by Aldrich and Connell detail the political, social and economic circumstances of overseas territories, most of which are islands. The British government's White Paper considers reform to the governance of some of these places. Dzurek details one of the most significant contemporary contestations over islands. Cruikshank's book is a study of an island group taken in war.

Aldrich, R. and Connell, J. (1992) *France's overseas frontier: Départements et Territoires d'Outre-Mer*, Cambridge University Press: Cambridge.

Aldrich, R. and Connell, J. (1998) *The last colonies*, Cambridge University Press: Cambridge.

Barbedo de Magalhães, A. (1992) *East Timor: Indonesian occupation and genocide*, Opporto University: Porto.

Blouet, B. (1989) *The story of Malta*, Progress Press: Malta.

Burnett, J. (1990) 'Birds know no boundaries: puffins and terns rule the roost on Machias Seal Island', *Canadian Geographic*, 110, pp. 32–40.

Cameron, J. (1992) 'The federated States of Micronesia: is there a Pacific way to avoid a MIRAB society?', in H.M. Hintjens and M.D.D. Newitt (eds) *The political economy of small tropical islands: the importance of being small*, University of Exeter Press: Exeter, pp. 150–79.

Christopher, A.J. (1999) 'New states in a new millennium', *Area*, 31, 4, pp. 327–34.

Connell, J. (1988) 'New Caledonia: a crisis of decolonization in the South Pacific', *The Round Table*, 305, pp. 53–66.

Connell, J. (1994) 'Britain's Caribbean colonies: the end of the era of decolonisation', *The Journal of Commonwealth and Comparative Politics*, 32, 1, pp. 87–106.

Connell, J. (1997) *Papua New Guinea: the struggle for development*, Routledge: London.

Cruickshank, C. (1975) *The German occupation of the Channel Islands*, Guernsey Press: Guernsey.

Daws, G. (1968) *Shoal of time: a history of the Hawaiian Islands*, University of Hawaii Press: Honolulu.

Department of the Taoiseach (1996) *Report of the interdepartmental co-ordinating committee on island development: a strategic framework for developing the offshore islands of Ireland*, The Stationery Office: Dublin.

Desmond, E.W. (1995a) 'Yankee go home', *Time*, 6 November 1995.

Desmond, E.W. (1995b) 'The outrage on Okinawa', *Time*, 2 October 1995.

Dodds, K. (1998) 'Enframing the Falklands: landscape, identity and the 1982 South Atlantic war', *Society and Space*, 16, pp. 753–6.

Dzurek, D. (1996) *The Spratly Islands dispute: who's on first?*, International Boundaries Research Unit, University of Durham: Durham.

Evans, J. (1992) 'White gold in the South Pacific', *Independent Magazine*, 28 March 1992, p. 18.

Falklands Islands Government (1999) *Statement by the Falkland Islands Legislative Council*, 14 July 1999, Falkland Islands Government Office: London.

Fathers, M. (1992) 'A tropical gangland', *Independent Magazine*, 25 January 1992, pp. 22–5.

Ferguson, J. (1998) 'A case of bananas', *Geographical*, LXX, 1, pp. 49–52.

Foreign and Commonwealth Office (1999) *A partnership for progress and prosperity: Britain and the Overseas Territories*, HMSO: London.

Gibney, F. Jnr (1996) 'Eviction notice', *Time*, 23 September 1996.

Goebel, J. (1927) *The struggle for the Falkland Islands*, Yale University Press: New Haven (re-issued 1982).

Haller-Trost, R. (1995) *The territorial dispute between Indonesia and Malaysia over Pulau Sipadan and Pulau Ligitan in the Celebes Sea: a study in international law*, International Boundaries Research Unit, University of Durham: Durham.

Hornblower, M. (1991) 'Corsica: let the stranger beware', *Time*, 18 February 1991.

Kunii, I.M. (1996) 'Yankees, get lost!', *Time*, 9 September 1996.

McCarthy, T. (1999) 'Reef wars', *Time*, 6 March 1999.

McElroy, J.L. (1998) 'A propensity for dependence', *Islander*, 5, 1998, p. 33.

Philpott, R.A. (1996) 'New light on the early settlement of Port Louis, East Falkland, the French occupation, 1764–7', *The Falkland Islands Journal*, 6, 5, pp. 54–67.

Ricketts, P. (1986) 'Geography and international law: the case of the 1984 Gulf of Maine boundary dispute', *Canadian Geographer*, 30, 3, pp. 194–205.

Royle, S.A. (1985) 'The Falkland Islands 1833–1876: the establishment of a colony', *Geographical Journal*, 151, pp. 204–14.

Royle, S.A. (1986) 'A dispersed pressure group: *Comhdháil na nOileán*, the Federation of the Islands of Ireland', *Irish Geography*, 19, pp. 92–5.

Royle, S.A. (1992) 'The small island as colony', *Caribbean Geography*, 3, pp. 261–9.

Smith, R.W. and Thomas, B.L. (1998) *Island disputes and the Law of the Sea: an examination of sovereignty and delimitation disputes*, International Boundaries Research Unit, University of Durham: Durham.

Strange, I. (1985) *The Falkland Islands*, David and Charles: Newton Abbot (3rd edition).

Walsh, J. (1992a) 'China pushes its weight around', *Time*, 6 March 1995.

Walsh, J. (1992b) 'Sea of troubles', *Time*, 27 July 1992.

8 Making a living
Island economies in the contemporary period

Resource dependency

A large country will normally present its inhabitants with a wide range of resources. There may well be variations in geology with a range of mineral resources available thereby. Such a country may well intersect with a number of climatic zones, with a variety of agricultural opportunities thus being presented. The same might be true for soil types and so on. A small island, because of its limited size, might not have much variety in terms of climate or soils; it will be able to produce just whatever crops or products suit its restricted ecumene. It is true that some islands exhibit tremendous variation in terms of altitude, especially volcanic islands. Tenerife is an example here, only 2060 km² yet in its volcanic peak, Pico de Teide, it contains the highest mountain in Spain. Islands such as this do have a variety of climatic zones and, thus, a range of opportunities. However, their restricted size means that the absolute amount of produce from any one zone will be small, the island will not be a major producer, and its output will thus be at the mercy of market prices and conditions dominated by larger producers elsewhere. Sheep farmers in the Western Isles, say, have insufficient scale of production to be able to affect prevailing British prices for lamb or wool. To be able to compete at all, island producers may well have to focus output into the one or few products in which the island has some sort of comparative advantage: the 'all the eggs in one basket' scenario (as introduced in Chapter 3). If there is a downturn in the market for the product or, if there is a problem with supply caused, say, by disease, then the impact on the island economy can be extreme. In his book on small states, which principally focuses on Iceland, Ólafsson noted that many have obtained decent growth rates. However, he concluded his analysis with the warning that

> small states must be prepared for instability in export prices and incomes which results from their dependency on a few primary or semiprocessed export products.
>
> (1998, p. 86; see also Potter, 1993)

Such risks are not theoretical. In Canada, Prince Edward Island (PEI) specialises agriculturally in potatoes. In 1991 a virus affected the island's potato crop and the agricultural economy was badly hit. Even in 1998 the island was still striving to restore some of its lost markets; in January of that year it was the island's Premier no less, Pat Binns, who announced the re-opening of the market for PEI seed potatoes in Mexico.

An alternative problem to disease is resource depletion. Regard the Pacific island of Nauru, which had extensive phosphate deposits. However, almost a century of phosphate exploitation has left much of Nauru's surface like that of the moon. The central plateau has been mined out, leaving a landscape of coral pinnacles. Only a coastal strip c. 150 m wide is still capable of supporting vegetation. The island had wealth, but in the 1970s when phosphate prices peaked, much of it was squandered in grandiose investments abroad. Nor has wealth brought universal happiness to the people. With little possibilities for domestic food production, many rely on processed food and health problems, including obesity, have become common. In 1992 Nauru took Australia to the International Court of Justice because of the environmental damage, coupled with the fact that Australia paid below market price for the phosphate, and was awarded US$103 m in an out of court settlement in 1993. Nobody expects a complete restoration of the ravaged island (Richardson, 1993). Many Naurans live abroad. Resource depletion may not just relate to mining economies as will be seen in the next section.

The case of Newfoundland

The eastern Canadian island of Newfoundland had a resource-based economy in the early 1990s, which was devastated by having been too reliant upon one particular renewable product that was over-exploited beyond its ability to recover. The product was fish, cod to be precise. Ocean resources had been the key to the region's European settlement history and economy from the earliest days. Various European powers competed for these resources including, at first, Spanish, Portuguese and Basques. Red Bay, an isolated village at the end of a road that links just five settlements in southern Labrador and eastern Québec, was once a very significant European settlement. This was in the mid-sixteenth century when it was the centre of Basque whaling activities, focused around an offshore island, Saddle Island, which offered some defence. Significantly, the Basque interest faded once the stocks of whales were depleted. As in the rest of what became Eastern Canada's Atlantic Provinces, two European powers remained active long term in the Newfoundland area, namely the British and the French. Both exploited the fish stocks, but it was the British who made most use of the island of Newfoundland as they processed the fish (by drying and salting) on the land rather than aboard ship as the French tended to do. Newfoundland ended up as a British possession, separate from Canada until 1949 when Newfoundland and Labrador voted to join the Canadian confederation. France retains a presence in the area as her once extensive North American empire became reduced to two islands off southern Newfoundland, what is now the 'territorial collectivity' of Saint-Pierre et Miquelon. From these islands France controls her regional fishing interests.

Newfoundland itself is largely wilderness: it is an island of harsh climate with shallow soils, much bog, standing water and forests. Only a few favoured spots such as the Avalon and the Port au Port peninsulas sustain any major agricultural activity. The one city of any size is St John's, focused around a fine harbour on the Avalon Peninsula in the extreme east of the island. Elsewhere there are a series of resource towns, mostly in the interior: Gander has an airport, very significant in the early days of transatlantic flights as it was the closest North American airport to Europe. Stephenville was an American airbase; Deer Lake grew up around a hydro-electricity plant; Grand Falls-Windsor and Corner Brook are papermaking towns (Norcliffe, 1999). Most other Newfoundland settlements are coastal and dependent upon fishing. Some, such as Bonavista and Grand Bank are fairly large, with populations in the thousands. Grand Bank still displays some splendid Victorian architecture as a souvenir of the riches once to be made from the seas around the island. Most other settlements are tiny fishing villages known as outports, with populations in the hundreds (Figure 8.1). The outports were totally focused on the sea; many were situated on offshore islands such as Fogo and the Change Islands to get better access to the fishing grounds. Some on the coast of the main island were never connected up by road. Newfoundland, even today, has a skeletal road network, basically the TransCanada Highway, which crosses the island from St John's to Channel Port aux Basques in the extreme southwest, which is the ferry port to Cape Breton Island, Nova Scotia and from there into the North American mainland. From the TransCanada, a few spur roads go off giving access to most but not all of the small coastal towns and outports. Those not connected, and those on the off-islands, were and are dependent, at least for the first leg of the journey to the rest of the world, upon boats and ferries. There was a railway, too, from 1898 but this closed in 1988. Its route was similar to that of the TransCanada Highway.

The small bays around which outports clustered could only house a limited number of boats. Further, the outports had to be within a few kilometres of exploitable fishing grounds or it became uneconomic in terms of time to reach the fishing area. As local population grew and pressure on shore space or offshore resources increased, new outports had to be founded. At the peak, there were around 1000 of these fishing villages. The outport system was never stable in that, should local opportunities decline, people also had to move. Between 1949 and 1954 almost 50 outports were abandoned by their inhabitants. In 1953 the Newfoundland government became involved in such resettlements and helped to relocate another 110 communities. From 1967 to 1975, government assistance became formalised as the Newfoundland Resettlement Programme and 150 more communities were helped, with resettlement being focused into 77 outports identified as possible growth centres as they offered more social and economic opportunities. Between 1965 and 1972, a total of 19,197 people were resettled at a cost to the Federal government of over Cdn$6.5 m with another Cdn$3 m coming from the provincial government. The programme was not altogether successful in that some of the economic

Figure 8.1 Long Island, an outport in Newfoundland, Canada, 1995

opportunities in the growth centres proved to be ephemeral and some of the people resettled actually moved back to their old homes. Rowe summed up the resettlement programme in the words of an anonymous fishing character: 'we was drove' (1985, p. 119): powerlessness again. This modernisation programme was characteristic of the then Newfoundland premier, Joseph Smallwood, who had pledged to drag his province kicking and screaming into the twentieth century (Rowe, 1985; Matthews, 1978–9). What Smallwood could not do was to successfully reduce the dependence of the outports upon the sea, which meant principally upon cod. Sadly the regional cod stocks were becoming depleted (MacKenzie, 1995). European fleets overfished the area; upon Canada declaring a 200 mile (320 km) limit around Newfoundland, Canadian fleets (and the French from Saint-Pierre et Miquelon) overfished the area. In 1992 the situation had reached such a serious level that the Canadian federal government, to try to allow the stocks of cod to recover, had to declare a two-year moratorium on catching cod: 'fishing gone' was the too lighthearted quip atop *Time*'s account of the story (Fedarko, 1993). Norcliffe's recent hard-hitting article (1999) is free from quips, using terms like 'destructive', 'exploitation' and 'mining' to describe the use and abuse of Newfoundland's resources: 'an orgy of high-tech fishing' is his description of the last days of the cod fishery (p. 107). The cod stocks did not recover; the moratorium was extended and at the time of writing, several years on from the first declaration, commercial cod fishing is still not permitted in Newfoundland waters. There is some exploitation still of other species, some of the processing plants limp on, working also with imported fish. However, most of the c. 25,000 fishers and the shore workers once dependent upon cod were rendered unemployed and they became dependent instead upon welfare, principally TAGS (The Atlantic Groundfish Scheme) payments.

Newfoundland, sadly, was the classic case of an island economy being too dependent upon a single product. The product failed; misery and a near shut-down of large sections of the economy resulted. Newfoundland strives to diversify, to find alternatives. The knowledge economy is one sector being developed. Tourism is another, capitalising upon the survival of the wilderness that resulted from Newfoundlanders turning away from the land; as well as upon the rich and distinctive cultural heritage donated to the island by its iso-lation, expressed in fine style in its traditional music. One newly important sector is the oil industry, associated with the Hibernia field off the east coast. By 1999 Hibernia-related activity was contributing about 5 per cent of provin-cial GDP, as opposed to 3.8 per cent from fishing and fish processing. Hibernia also has beneficial effects in that many of the people trained to work in the oil and associated construction industries gained transferable skills, including those in high-tech activities. The oil industry is a useful diversifica-tion for Newfoundland, but the province remains the poorest in Canada. Nor-cliffe (1999) pointed out something of the social costs of Newfoundland's crisis, with Newfoundland having Canada's lowest birth rate and highest out-migration rate by 1998.

Sequential resource use

If an island loses an economy on which it relied, alternatives have to be found, as with Newfoundland's struggle to diversify from cod as described above. If alternatives do not emerge, there is the likelihood of population loss. Some islands have managed to develop a series of sequential economies bringing dif-ferent resources on stream in turn, and two examples will be discussed below.

The case of the Falkland Islands

Ship resupplying and then sheep farming supported the Falkland Islands' economy into the 1980s, as explained in Chapter 3. At that time, the sheep ranches were suffering from underinvestment and their output of wool was under challenge from synthetic fibres. In 1987 the Falkland Islands government started to license fishing within an exclusive economic zone. Anybody wishing to fish within this zone – the principle catches are illex and loligo gahi squid – must purchase a licence from the Falklands government. On this activity the economy of the Falkland Islands currently depends; in recent years over 50 per cent of the government's operating revenue has come from the fisheries sector. Much of the recent development in the islands, such as the provision of the £14 m community school, which includes a public library and sports centre, has been funded by the local government and not by the UK. The economic focus of the Falklands has shifted once more to Stanley, which has been growing in recent years, not least because of the exodus of agricultural labour from Camp.

The Falkland Islanders are conscious of the fragility of their reliance upon the selling of licences to fish to Spaniards, Japanese and South Koreans as the basis of an economy and the constant fear is that overfishing or illegal fishing will harm the stocks. To counter these problems, the Falkland Islands took on *M.V. Dorada* as an extra fisheries protection vessel in 1997. In 1999 the islands' Chief Executive, concerned with poaching from Taiwan, reported that local boats were being armed (Andrew Gurr in the *Independent on Sunday*, 9 May 1999). A further problem is there is fluctuation in earnings year on year: fisheries revenue fell from £23 m to £18.3 m between 1996–7 and 1997–8, for example.

There were hopes that a fresh sequential resource would be found to ease the problems of over-reliance upon fish licensing. This was oil, the Falklands being part of a sedimentary basin. There are only 2000 Falklanders and it would not have taken much of a find to make the islanders very wealthy. However, at the time of writing exploratory drilling has not so far discovered exploitable reserves, and oil remains only a possibility for the next major contributor to the Falklands' economy. In the meantime the Governor stated there would have to be 'a more prudent and rigorous approach to financial planning' and to deplete reserves built up from fishing income would be 'highly irresponsible' (Falkland Islands Government, 1998, p. 5). There is also a programme of diversification; that is the mission of the Falkland Islands Development Corporation, helped by the fact that there is certainly much greater entrepreneurialism abroad on the islands now than in the period before the Conflict. In the 1997–8 financial year the Corporation was working on the development of calcified seaweed as fertiliser, the establishment of a beef herd, and the building of a proper abattoir, to effect better marketing possibilities for island meat (Falkland Islands Government, 1998). There has been the development of the 'Falklander' sweater to try and stimulate value-added manufacturing on the islands instead of the export of unprocessed wool. Tourism is being encouraged; both from cruise ships and land-based tourists, though for the latter it is a difficult and expensive journey to reach the islands. In 2000, the abattoir had been completed, allowing meat to be exported from the Falklands, and the Department of Agriculture was beginning the process to have the island's entire output certified as organic. The cool climate means that pesticides are not required, and the farmers with their huge holdings had never taken up what would once have been seen as modern – intensive and non-organic – methods.

The case of the Turks and Caicos Islands

Another example of sequential economies comes from the Turks and Caicos Islands. The first economy was based on South Caicos, Salt Cay and Grand Turk, that were used to produce sea salt. These islands are low lying and were partially excavated to produce salinas, a grid of bunds being built to form inter-connected, flat floored, shallow basins, which could also be isolated from each other. The sea was allowed access to these salinas and ponded behind the bunds. The region's strong sunlight, together with the wind whipping across the flat, treeless islands made ideal conditions for rapid evaporation. Sea salt was precipitated out and would be raked up (Grand Turk's principal hotel is the Saltraker Inn). The salt from the Turks and Caicos Islands was then traded, mostly up to the Canadian fishing grounds where it was used to salt, i.e. preserve, the fish. This exploitation of the islands was initially controlled from Bermuda and this put them within the British sphere of influence. They remain a British Overseas Territory. The three salt islands, the two Turks Islands and just one of the five Caicos Islands, were the engines of the economy. Grand Turk houses Cockburn Town, the principal settlement with the wide range of activities and functions of a self-governing colony. The other Caicos Islands were barely inhabited, just a few fishing families being based on them.

The salt industry collapsed in the 1960s, techniques for preserving fish, especially canning and freezing, replaced salting to a large extent, the demand for the Turks and Caicos' principal product dropped and the salinas ceased to operate. This left the islands bereft. Salt Cay is now a UNESCO World Heritage Site, which celebrates its legacy from the industry but has never recovered economically from the closure of the salt works. Grand Turk fared better, partly because it was the centre of administration for the colony, but also because it had become the site of two American bases, prosaically called North Base and South Base. The latter was focused around the airstrip, which had been built by American forces during World War II. The bases provided much local employment and replaced, for a period, salt production as the mainstay of the archipelago's economy. This made Grand Turk ever more dominant in the islands. The bases closed in the 1970s, leaving in the case of North Base, a derelict landscape to add to the unsightly disused salinas. South Base, by the airport, was reused as government offices.

Fortunately, further functions have now, to some extent, rescued the Turks and Caicos Islands. One element is offshore finance, which provides about 300 jobs and 7–8 per cent of GDP, but this has not been free from problems as will be seen later in this chapter. More important is tourism, which provides 45 per cent of GDP. The islands are warm with fine beaches and, most significantly, they are an easy 75 minutes flight from Miami. Tourism has recast the settlement geography of the islands, for it is the Caicos Islands – under-used for a long time – that have been taken over by the new economy. In particular Providenciales (normally called Provo) has replaced Grand Turk as the engine of the islands' economy. This island has had a service centre, Downtown, with a full range of retail and other services, erected close to its airport of 1984. Along the island's coastline from the 1980s a series of hotels have been built for multinational companies. Wharves and other marine facilities have been erected to take advantage of the island's fine diving sites, and many holidaymakers take boats to go diving and/or to go sports fishing. A golf course has been built, a startlingly artificial green area in the island's otherwise olive-coloured, scrubby landscape. Provo has had much immigration from islanders wishing to avail themselves of jobs in construction and the tourist industry and is now the most populated island with an estimated 15,000 people in 1999, compared to around 1000 in 1986. Its population and economic dominance is a complete reversal of the situation during the Turks and Caicos Islands' other economies. Some of the residents of Provo are illegal migrants from Haiti and Haitian patois can be heard in some of the poor quality housing areas along the spine of the island, away from the hotels on the beaches. Haitians

> come by plane, they come with visas, but mostly they come by boat in the middle of the night. They come to seek a future and to send money back home to help those they left behind.
>
> (*The Turks and Caicos Times*, 25 April 1999)

In 1999 the government adopted a tougher stance towards the Haitian shanty settlements, which present an unwelcome image of the islands, and began making raids on them. A local newspaper reported with some relish on the 'stench of human excrement, rotting garbage, fly infested corners and mounds of debris' in Clement Yard on Provo, occupied by 15 families and within walking distance of the Beaches Resort and Spa. The Haitian man who owned the properties was given instructions to make improvements and the Assistant Director of Planning promised more raids on similar settlements once his department was 'beefed up with staff' (*Turks and Caicos Weekly News*, 15 April 1999).

It is necessary, of course, for the Turks and Caicos Islands, whose economy now depends on yet another single industry, to continue to invest in tourism development to maintain market share in what is a fickle business (see Chapter 9). The latest proposal is to bring into play the still barely populated East Caicos, much of which is slated to be given over to facilities for cruise ship tourism. Facilities are planned for up to 660 cruise liner visits per year, with 1.6 m tourists, and would involve nearly half of this island going under the bull-dozer. Perhaps 7000 jobs would be created and the archipelago's future secured thereby. In fact there would inevitably be a labour shortage and there would have to be considerable immigration, a process not necessarily free from social problems. The potential developer, Bill Grenier of Pagebrook Inc., was sub-jected to a considerable grilling in the local press on the proposal, which appeared over several editions of the *Turks and Caicos Weekly News* in early 1999. The project's large scale has led to concerns being expressed about its impact on the environment, particularly on the habitats of the roseate flamingo, iguanas and turtles, as well as vegetation in the salt marshes and mangrove swamps. However, in economic terms it may well be considered that tourism is now so vital to the local economy that some loss of habitat on one of the islands is a necessary sacrifice. In his interview, the developer stressed that

> of course we are not going to destroy the island, the island and its environ-ment is exactly what we are trying to promote.
>
> (*Turks and Caicos Weekly News*, 15 April 1999)

The basic point of the development was to scale up the tourism industry for economic benefits:

> One of the objectives of the port is to bring tourists who will leave sufficient money behind ... to enable the port to pay its way and make money – not one airplane at a time, but one shipload at a time.
> (*Turks and Caicos Weekly News*, 15 April 1999; see also the *Independent*, 5 December 1998)

We see in the very different examples of the Turks and Caicos Islands and the Falkland Islands, islands that have had the advantage of having been able to uncover a series of economies over the years. However, in both cases there is a concern that the current engines of the economy may prove to be unreliable even in the medium term. Small islands such as these almost inevitably struggle to develop a comfortable long-term economic position. Neither group has yet become strong enough to become independent of the UK, although, especially with the Falklands, there are other issues than the economy to be considered regarding independence.

Coping with scale problems: scaling up

Islanders have adapted different strategies to try to cope with their economic problems. One is to try to scale up production and services. Thus British West Indies Airways service the English-speaking Caribbean; Air Marshall Islands and Air Nauru provide international air links not just for their own islands but for other island nations such as Kiribati and Tuvalu. The Trans-Balear shipping company services the shipping needs of all Spain's Balearic Islands. Caledonian MacBrayne operates most of the ferry services to the Hebrides and Clyde islands of Scotland as explained in Chapter 6. Scaling up can be seen in other activities, too. The West Indies cricket team is one example from sport; from education examples are the University of the West Indies and the University of the South Pacific. In trade and politics, there are Caricom, the Caribbean trading organisation, and the South Pacific Forum, which is a regional political organisation, amongst others.

However, in this chapter the policy must be discussed in the economic sphere and the best example is in the development of island co-operatives.

Island co-operatives

These organizations are to be found on almost every small Irish and Scottish island, for example. The idea is that producers combine together to maximise the advantages of economies of scale: if all the farmers put in a joint, bulk order for feed, the unit cost to individual farmers will be less than if they had each made small individual purchases. The co-operative can organise more efficient marketing and can purchase conjointly expensive equipment individual farmers could not afford or justify themselves, but which as co-operative members they can use on a rota basis. There are island co-operatives for specialist local products, thus there is a textile co-operative on Scalpay, one of Scotland's Western Isles, but often on small islands the co-operative is a general, multifunctional business. Inishmaan is the central island of the three Aran Islands in Ireland's County Galway with a population of 216 at the 1991 census. Its co-operative, *Comharchumman Inis Meain Teo*, is a multi-functional business with an annual turnover measured in hundreds of thousands of pounds. It was founded with government encouragement in the 1970s. Soon after its inception, the founder said:

> Two years ago the death of the island was being forecast. It was a long time since there was a marriage in the place or since anyone returned.... It was the only ... Aran island ... without a water ... or electricity scheme.... Within the last year there were three weddings ... and five households returned.... We are only beginning. We have plans for tourist, trading and transport projects.
>
> (cited in Johnson, 1979, p. 71)

This co-operative has had the mission, the will and, most importantly, the resources from the scaling up of those of the individual members, to have been able to invest in infrastructural improvements of benefit to all. These have included an electricity scheme, the renovation of the community hall, the building and installation of piped water, a summer college scheme (to teach Irish to paying visitors), a football pitch, a telephone exchange, telecommunications facilities, and the tarmacking of the island's airstrip. One other notable development was a textile factory, established in 1976 in a shed. This employs about 15 people, a large proportion of the workforce, now in purpose-built premises. Its output, of course, is small and the business would be a tiny, disadvantaged player if competing for the mass market. However, sensibly, its policy is instead to trade in the luxury market. Thus the returns from its small output are maximised and the business tries to establish a brand name, with a reputation which will bring repeat business: these are not just sweaters, but Inishmaan sweaters. The marketing makes great play of the romance of the island's isolation and heritage, including its being Irish speaking. The author bought a sweater there on his last visit, the bag and label declared that these sweaters were traded in London, Paris, Tokyo and the Aran Islands. This is an example of island branding which will be discussed further below.

Remaining with Ireland, the various island co-operatives were active in the instigation of an all-Ireland island pressure group, the *Comhdháil na nOiléan* and its work which helped to initiate the process that resulted in an Irish government ministry being charged with overseeing island affairs, as explained in Chapter 7.

Coping with scale problems: scaling down

As explained in Chapter 3, islanders are often forced to engage in occupational pluralism to exploit their limited resource base. Islanders engage in a series of part-time jobs that, in other places, would each occupy people full time. Each of the jobs is thus scaled down for the island situation. The people of Tristan da Cunha work the land, work at their service jobs, whatever, but many of them drop their normal occupations when the sea state is such that a 'boat day' can be declared. Then many able-bodied male islanders become fishermen and afterwards most Tristanians help to process the catch in the island's fish factory. Most days are not boat days and islanders revert to their other occupations and activities. There are not enough islanders and not enough boat days – or fish – for there to be professional, full-time fishers, so fishing, though of overwhelming importance to the island economy is carried out in a scaled down, part-time fashion.

Scaling down in occupational terms is also seen with regard to service jobs. A prime example was the career of one Seamus McSporran of the small inner Hebrides island of Gigha. McSporran held no fewer than 14 jobs, all part-time, all of which in a larger mainland society would have been filled by full-time professionals, but which in Gigha have been scaled down to be just part of one multi-skilled man's duties. McSporran's activities included being shopkeeper, postmaster, petrol pump attendant, undertaker, ambulance driver, taxi driver, harbour master, constable and firefighter (Figure 8.2). He retired in 2000 and his remarkable career merited a mention in *Time* (24 April 2000).

Scaling down has limits, of course, and often small islands simply lack the scale to support any on-island provision of a particular facility, and arrangements have to be made to access it off the island. This can be related to everything from going off-island for shopping to being sent for medical treatment elsewhere.

Coping with scale: island branding

Branding is another well-tried strategy for island economies. If there is an identifiable brand or product, which is, or can be made to be, associated with an island, then its small output should carry a premium mark up and thus maximise returns. In 1986, 60 per cent of wine produced on the Portuguese island of Madeira was exported in bulk, mostly to France and Germany where it was used as an ingredient in cooking sauces. However, this island has a brand identity as the home of Madeira, a variety of wines made from special grapes that have also undergone *estufagem*, a unique 'baking' process within the casks, which are heated. This process produces the taste that characterises this expensive product and is itself a result of Madeira's islandness; *estufagem* was developed as it helped to preserve the wine during the long sea voyages necessary to take it from the island to market. Thus for Madeiran vine growers to produce output from hybrid grapes that ended up in German sauces was a wasted opportunity. The Madeira Wine Institute, an arm of the island's government, wanted to replace this low quality output by increasing that of branded Madeira from 3.5 m litres to 5 m litres annually and also export more product in bottles, rather than in bulk, which would also bring added value to the island.

Figure 8.2 The business premises of James (Seamus) McSporran, Gigha Island, Scotland, UK, 1999

Another example relates to Channel Islands sweaters. These are not just any old sweaters, but the original Jerseys and Guernseys (sweaters are known in some places as 'ganseys', a corruption of 'Guernsey'). There are also Aran sweaters named from the Irish Aran Islands. Off-island producers can only make Jersey, Guernsey and Aran-style sweaters; on-island producers supply the genuine articles. Thus producers on Guernsey and Alderney, an island in the Bailiwick of Guernsey sell – for premium prices – branded Guernseys, the traditional garment of their islands made since the sixteenth century from oiled wool and to a distinctive design. Prices on the Internet currently range from US$95 to $145 per garment, dependent upon size. The 'Falklander' sweater was mentioned above. This is not a traditional garment at all – its design was actually the subject of a competition on the BBC television programme, *The Clothes Show* – the hope is that it, too, can establish itself as a branded product and thus, too, command premium prices.

Weavers on the Isles of Harris and Lewis make not just cloth, but Harris tweed, 'originally developed by the islanders to withstand the stinging rain and gales driven in by the wild Atlantic' (MacGillivray and Company, 1989). It is marketed with one of the world's oldest trademarks and quality control labels – the famous orb symbol – and thus tweedmakers elsewhere cannot make Harris tweed.

Other Scottish islands specialise in the production of branded whisky or liqueurs. Drambuie comes from the Isle of Skye; Bonnie Prince Charlie and all of the island's romantic past is exploited to sell more product. Some way south of Skye lies the island of Islay. It produces whisky, not any common or garden variety but the famous Islay single malts: Laphroaig, Bowmore, Ardbeg, Lagavulin, Caol Ila, Bunnahabhain and Bruichladdich. These brands sell at premium prices as specialised niche products; thus there was an article on Ardbeg in the British newspaper most likely to be read by those with high incomes, *The Financial Times* (30 August 1998). Laphroaig is the most heavily advertised and makes much of its reputation as the most peaty Scotch, 'as the malted barley is dried over a peat fire. The smoke from this peat, found only on the island of Islay, gives Laphroaig its particularly rich flavour', as it says on the Laphroaig bottle. If you step up to your local bar and just ask for whisky, you will not get an expensive Laphroaig. It and the other Islay malts are named brands; you ask for them specifically. Islands noted for named, branded products of course also use their reputation to encourage tourism; many go to Islay partly to visit the distilleries. Back to scaling up, the seven Islay distilleries market themselves, together with the Isle of Jura distillery on the neighbouring island as 'The Islay and Jura Whisky Trail'. Another Scottish island is reviving distilling to try to break into the branded malt whisky market. This is the Isle of Arran where a distillery opened in 1995. First deliveries of Isle of Arran Single Malt, 'the new spirit of Scotland', were not due until 2001; to tide it over until that time, the distillery sold blended whiskies and marketed malts made elsewhere.

Offshore finance

In the late twentieth century banking and finance became a global business. The liberalisation of international capital flows from the 1960s, especially the removal of fixed exchange rate mechanisms, and the development of computing and telecommunications were important factors in this development. Some small self-governing islands play a major role in the banking and finance industries, hence the general use of the terms 'offshore banking' and 'offshore financial centres' (OFCs) (Hampton, 1996; Hampton and Abbott, 1999). These islands capitalise upon any comparative advantages that they can establish through their legislative independence, such as the ability to establish favourable fiscal regulations that make it financially advantageous for wealthy individuals and companies to use the island, for example to reduce their tax burden. A location conveniently accessible to prosperous areas and wealthy people helps as well. Thus the Channel Islands and the Isle of Man offer conveniently situated offshore financial services for Britain and Northern Europe. The Isle of Man Bank points out that having been founded in 1865, it is the first entry in the Isle of Man Register of Companies. This gives the bank stability and tradition, but the Isle of Man Bank's marketing is also based on insular characteristics. These include the 'thousand years of economic and political stability created by the Manx government, Tynwald' (founded in AD 979), Tynwald's 'freedom to legislate in respect of domestic matters', and the island's separation from the fiscal regulations of the UK which has enabled it to 'develop a tradition of low taxation' (Isle of Man Bank, n.d.).

Small islands thus strive to become attractive OFCs, but, regarding Jersey, another OFC in the British Isles, Hampton and Christensen (both Jerseymen, incidentally) rather wonder if the small island situation is not exploited by outside actors – island powerlessness again.

> Jersey's apparent success as an OFC is not based purely on low taxation or strict bank confidentiality (in effect, bank secrecy), but appears to exist with British political acquiescence and local connivance. Contrary to official versions of the island's history (States of Jersey, 1993), it appears to be a case of rather more luck than judgement: the case of a small economy being 'discovered' then increasingly exploited by international financial capital since the 1960s, rather than a planned state-led development to become a finance centre.
>
> (1999, p. 1620)

Hampton and Christensen note that Jersey's OFC is the largest contributor to the island's GDP and also provides jobs for about 20 per cent of the workforce. As always with small islands, overdependence upon a single industry is risky and the authors go on to analyse a series of problems that arose in Jersey's finance sector in 1996 regarding a US$26 m fraud, contentious legislation on limited liability partnerships in the island's parliament and a political crisis. Their conclusion is that even in this well-established and internationally recognised OFC there was 'regulatory failure' (p. 1619). This contributed to a growing feeling within Western democracies for:

> accountability, transparency and probity in public life, [and] it is unclear how much longer offshore finance centres can continue to exist, even in highly modified form. Even more questionable is the creation of new OFCs by small island hosts who aspire to be so-called 'international financial centres'.
>
> (Hampton and Christensen, 1999, p. 1635)

It is now widely recognised that not only is it unlikely that there will be new entrants to the market, but already there are too many OFCs. Some of the more recent entrants have struggled to establish market share, given competition from longer established centres. Excess competition or perhaps simply greed may explain why some jurisdictions have engaged in illicit practices and there have been many accusations of unscrupulous investors using some OFCs for money laundering. 'A trillion dollars in dirty money keeps island tax havens afloat' was one headline that left the reader in little doubt that probity was not always uppermost in all OFCs (*Independent*, 11 September 1996). The accompanying article was careful to note that not all such centres were necessarily shady, although the offshore finance industry generally has attracted criticism in recent years for being often poorly supervised and regulated (see, for example, Aldrich and Connell, 1998). The international community has become increasingly interventionist with regard to offshore finance. The United Nations in 1999 set up the Offshore Forum to consider how best to regulate the industry. A year earlier the Organisation for Economic Co-operation and Development (OECD) began to deal with unfair tax competition. In 2000, the organisation published a blacklist of 15 states and territories where money laundering was found to take place (Financial Action Task Force on Money Laundering, 2000). Although one newspaper report chose to headline that 'Russia and Israel are criminals' cash havens' (*Independent*, 23 June 2000), in fact they were amongst only five continental states on the blacklist (the others were Lebanon, Liechtenstein and Panama). The rest were islands: the Bahamas, Cayman Islands, Cook Islands, Dominica, Marshall Islands, Nauru, Niue, Philippines, St Kitts and Nevis and St Vincent and the Grenadines. The newspaper report added that the Channel Islands and the Isle of Man were included on a 'grey list' of other OFCs under surveillance. At the European level, the EU is moving forwards on tax harmonisation including plans for a withholding tax. This is of potential difficulty to the three offshore centres in the British Isles: Guernsey, Jersey and the Isle of Man. A senator on Jersey is drawing up plans to hold a referendum on independence if the EU demand that the UK change the island's fiscal regulations.

The North Americans are also concerned about money laundering in the Caribbean. About a dozen members of the Royal Canadian Mounted Police accompanied members of the Turks and Caicos Islands' police force in a raid on an offshore bank's offices in Grand Turk in February 1999 on suspicion of money laundering. Island lawyers managed to stop the Canadians removing from Grand Turk boxes of documents, whilst island bankers noted that the Royal Turks and Caicos Police Force would hardly have been able to operate in a similar fashion in Canada – island powerlessness (*Turks and Caicos Weekly News*, 15 April 1999). Advertising material in 2000 from the Turks and Caicos Islands stresses the new role its Financial Services Commission (FSC) now has in regulating the industry:

> The FSC supervises the entire offshore sector, which includes banking, insurance, trust and investment activities and company registration. Under new legislation the FSC is about to start licensing company managers, agents and mutual fund administrators. The FSC formerly undertook marketing of TCI [the Turks and Caicos Islands] as a financial services center, but its role is now purely regulatory. . . . The stability and confidence provided by the fact that TCI is a UK Overseas Territory is another pull factor. The recent appointment by the British and island governments of a team to review regulatory standards and laws is warmly welcomed.
>
> (International Media Corporation, 2000, p. 2)

One message from this section is that the market for OFCs is now probably closing against new entrants, whilst some established players struggle to repair their reputations. One of the most successful established OFCs is Bermuda, which benefited from early development of expertise in certain specialised sectors of the market, particularly within insurance. Bermuda now plays on its experience and expertise, as well as its stable social, political and economic systems (Royle, 1998). In 1995 its electorate rejected independence by about 3:1 in a referendum and some reports opined that their decision was for reasons of 'financial security' (*Independent*, 18 August 1995).

Islands and the global labour market

Other islands have made their way in the contemporary world by participating in what became known as the new international division of labour. This is an aspect of globalisation which sees transnational companies seek to minimise costs and, thus, maximise returns by seeking optimal locations for each of the different processes involved in their operations. Thus, research and development work need not be carried out in the same place, even in the same country, as the assembly and manufacture of their products. Countries that offer cheap labour may well be chosen as locations for those aspects of the process that require heavy labour inputs. This affects islands insofar as some have been able to offer cheap labour and have thus been able to join this new international division of labour. This is not something in which tiny islands can participate, but several larger islands have become major players.

From the 1950s Taiwan was able to begin to industrialise on the basis of cheap labour and developed a reputation for its inexpensive, perhaps even shoddy products. This island has now left this early reputation behind and has become a major player in the electronics industry. More recently, islands undergoing industrialisation have tended to be linked into international companies, rather than rely on domestic enterprises. Islands capitalising upon their cheap labour are competing against Malaysia, even against China in today's globalised world. However, as long as the island can offer a sufficiency of labour with adequate skills, it may be able to so compete. Its islandness can even be of positive benefit as will be demonstrated in the examples of Mauritius and Batam Island, Indonesia, which follow below.

The case of Mauritius

The Indian Ocean island of Mauritius, because of its typically insular political and economic history, has been left with a range of ethnic groups, speaking, collectively, a number of major European and Asian languages. English is the language of government, French the lingua franca, and citizens originating from the Indian sub-continent and China also have Asian languages. Multilingualism within a workforce is an attraction to companies trading globally. Further, Mauritius' complex history has left the island with a clear majority of Hindus. Voting tends to be along ethnic lines and so a Hindu-dominated party has governed the island since independence. This has given Mauritius a stability of governance which has been completely absent from other African countries

(Mauritius is an 'African' island) and it is the only African state to have been free of major political strife since independence. Political stability – however obtained – is attractive to international business. Mauritius has been able to develop a good educational system and its people are reasonably educated with high levels of literacy. Social change from the 1950s and 1960s saw women becoming increasingly interested in taking work outside the home and this boosted the available workforce with people who were prepared to labour for relatively small remuneration. Finally, another typical facet of insularity saw Mauritius suffer a classic crisis in its population: resource ratio from the 1940s when deaths from malaria were considerably reduced. That, combined with a rise in the birth rate and total fertility rate saw population numbers swell to a point beyond which the island's traditional economy, then almost totally reliant upon sugar production, would have been able to offer support. The island authorities, at first the colonial British government, later the island's independent government, reacted in two ways. One was to try to reduce the birth rate by instigating an aggressive and well-organized family planning programme. The second was to encourage economic diversification.

Diversification took three main forms: into tourism, given the island's tropical location and splendid scenery; into offshore finance, linked to an extent into the island's association with the Indian sub-continent – an island as stepping stone touch here – and, finally, into manufacturing. Regarding the links with India, a double taxation avoidance treaty signed between the two countries in 1983, has been successful in attracting foreign investment into India through offshore facilities in Mauritius. 'Mauritius remains India's friend' was one recent headline on this operation (Bikoo, 1997, p. 2). Regarding manufacturing, the features listed above became important, features which were predicated upon insularity. Mauritius specialised in textiles, becoming a 'pyjama republic' in Huw Jones' memorable phrase (1989a). The island became an export processing zone, sending goods overseas that had been made on the island from raw materials not produced there; Mauritius supplied the premises and the labour. By 1997 the export processing zone was responsible for producing 69 per cent of the island's exports (Selwyn, 1983; Jones, 1989b; 1993; Alladin, 1993; Royle, 1995).

The case of Batam Island

Another island that came to participate fully in international manufacturing was Batam Island, Riau Province, Indonesia. Here it was not the labour reserves then present on the island that was the chief factor in its industrialisation. Rather it was the island's location that was the key. Batam is situated on the periphery of the huge Indonesian archipelago and just 21 km off the south coast of Singapore. This enabled Batam to participate in the regional growth triangle, which includes, in addition to Riau Province, Singapore and the Johor area of Malaysia. Batam's contribution is its availability of land, something in very short supply in Singapore, and also a favourable fiscal regime set up by

Indonesian authorities keen to see business attracted to Batam that could mop up some of the state's surplus labour. Singapore has excellent connections by air and sea, a sophisticated financial system and entrepreneurs keen to help bring business to the local region to the benefit in some way for Singapore's traders and financiers. It lacks land and cheap labour: one of these Batam could provide, the other it could obtain.

Before its industrialisation, Singaporeans were already using Batam as a holiday retreat. In the 1980s and 1990s, Batam's empty spaces began to fill with industrial premises. Three large industrial estates were set up, with other infrastructural investments in transportation and other services. The parks tended to be largely self-contained. The largest, Batamindo Industrial Park, had its own power supply, water and sewerage systems, shopping areas, leisure facilities, residential accommodation and mosque. Some Batam people worked there, but prior to industrialisation, the island's population was only about 155,000, suited to its then fairly basic rural economy. Thus labour had to be imported from elsewhere in Indonesia to perform the assembly work required by some of the world's major multinationals, companies such as AT&T, Philips, Varta and Ciba. The imported labour is housed in on-site residential blocks.

Batam's economic transformation has seen the transformation, too, of its traditional society and the complete reconstruction of much of its landscape; many areas not vested for building are planned to be given over to the development of reservoirs and other infrastructures. The population is slated to rise to 700,000. There has been little attention paid to environmental conservation, such was not always a prime concern during Indonesia's rush for modernisation in the late twentieth century. Batam's location, its peripherality, its boundedness – which allowed its development, including the supply of labour to be tightly controlled – its very insularity, have been the keys to its changed economy (Grundy and Perry, 1996; Royle, 1997; see also Mitchell, 1994, on the Indonesian development system as applied to another island, Bali). One general attraction to multinationals with regard to Indonesia in the late twentieth century was its political and social stability, despite some problems identified with corruption (Evers, 1995; Server, 1996). The stability was not always achieved in ways that met with the approval of western governments, in relation to civil liberties and human rights, but multinational businesses tended to take a more pragmatic attitude. The financial crisis in Asia in the late 1990s was accompanied in Indonesia by political unrest and law and order challenges, and also, sadly, ethnic violence in some islands. Such unstable phenomena are unpalatable to business investors. One commentary on Jakarta, the Indonesian capital, summed up the problems well in its title: 'From global city to city of crisis: Jakarta Metropolitan Region under economic turmoil' (Firman, 1999). Batam Island's extensive plans for continued industrial expansion were put at risk.

Self-support, outside help and MIRAB economies

Previous sections have shown how, in some circumstances, small islands have been able to utilise some of their qualities to be able to sustain themselves in the modern world. Many cannot fully sustain themselves and have to be helped in some way from outside. Often such help is in the form of national and regional development policies. Thus, on the island (and province) of Prince Edward Island (PEI) in eastern Canada, regional development policy has been applied by the federal and provincial governments for decades. This island is part of Atlantic Canada and thus is subject to the ministrations of the Atlantic Canada Opportunities Agency (ACOA), the latest in a line of development initiatives (Savoie, 1999). Over the years there has been a trend away from the earlier imposition of large-scale projects and initiatives imposed from the top-down, and the latest moves are to encourage bottom-up community economic development, utilising more of the islanders' own energies and self-reliance. The instigation of the island's four Rural Development Corporations (RDCs) was part of this transition. The RDCs facilitate development ideas emanating from community groups and individuals, down to the level of the achievement of self-employment by individuals. The favourably located Central Development Corporation, in whose area the Confederation Bridge makes landfall, mainly operates industrial units, although there are some tourism projects and the Corporation deals with its clients' individual needs, too. *La Societé de Devéloppement de la Baie Acadienne*, which operates in the more peripheral Evangeline Region, whose people are Acadian, and, thus, Francophone, is not able to do much with industry but has set up and operates tourism facilities, retail developments, offices, educational and IT projects. It would still like to open a golf course to attract more tourists, although this last scheme has failed yet to receive financial support from federal and provincial agencies, without which such developments rarely take place on PEI. At a world standard, PEI is wealthy, its GDP per capita in 1998 was Cdn$22,091. However, by Canadian standards this was low, at only 75.35 per cent of the national average, and PEI and the other major islands of Atlantic Canada, Cape Breton and Newfoundland, suffer relative poverty within their nation and it is thus deemed necessary to continue to support these island economies. Thus PEI received Cdn$400 m in federal government transfer payments in 1996, part of a total federal expenditure on the island of Cdn$1234 m. There was Cdn$631 m in federal revenues to offset this, but the balance of Cdn$603 m given a population total of 136,500, works out at per capita support of Cdn$4417.58 (Province of Prince Edward Island, 1999). On PEI and the other islands, the extensive regional development programmes do encourage self-reliance but they also require outside support – the Evangeline Region's golf course will not be built until outside funds are released.

Islands in less wealthy nations may well require, if not receive, comparable support. Many have been recognised as belonging to that category of places known as MIRAB economies (MIgration, Remittances, Aid and Bureaucracy).

The terminology, applied, for example, by Bertram and Watters (1985) to South Pacific microstates and by Aldrich and Connell (1998) specifically to Tokalau, Niue and the Cook Islands, is easy to explain. The islands struggle to sustain their populations, some of whom migrate as a result. Thus, a considerable proportion of South Pacific islanders live in Auckland, New Zealand; several hundred thousand people of West Indian descent live in the UK, etc. Migrants often send money back home to their families, these are the remittances and such transfers can form an important part of an island's economy (see Stanwix and Connell's (1995) study of remittances from Fijians living in Sydney). Aid comes from international agencies and/or bilaterally from former and, in some cases, from present, colonial rulers. The UN, for example, has had studies and programmes to aid island developing countries since the 1970s, and now, through UNCTAD (United Nations Conference on Trade and Development), focuses on the problems facing SIDS (Small Island Developing States) (Encontre, 1999). On the Marshall Islands during 1998, development projects were under way funded by the Asia Development Bank and also by the Japanese and US governments, these two nations both being former colonial rulers here. Bureaucracy refers to the situation in small islands for the state sector to be overwhelmingly dominant, there not being the scale or the skills in the private sector for it to carry out the roles that would be normally assigned to it in larger, more developed economies. Thus in Kiribati, the state in the mid-1990s owned and operated not just schools and libraries but transport systems, the hotel on South Tarawa, the broadcasting system, the bank, etc.

MIRAB economies are not confined to islands. Lesotho is probably more dependent upon remittances than many island states. In addition to its West Indian origin population, the UK has hundreds of thousands of migrants and their descendants from such major continental countries as India and Pakistan. However, the proportion of Indians and Pakistanis who emigrated is tiny compared to the proportion of Jamaicans, Trinidadians and Kittians, and the impact of remittances and the other MIRAB factors is obviously much greater proportionately in St Kitts than in India.

Conclusion

This chapter has demonstrated that small island economies often struggle to manage on their domestic resource base, although there are islands that have been fortunate to discover or develop sequential economies to support themselves. Others have had some success as OFCs, as participants in the globalised manufacturing process, peddling their reserves of cheap labour, or by establishing brand awareness for their products. Insular problems have been seen to loom large, however, and many islands need outside support. The next chapter considers in detail one near universal way in which islands in the modern era have tried to overcome some of the economic problems considered above. This is by engaging in tourism.

References

Key readings

On offshore finance Mark Hampton's book is a detailed study and the same author's paper with John Christensen on Jersey is a useful case example. Glen Norcliffe has written a hard-hitting condemnation of resource exploitation in Newfoundland. Huw Jones' work on Mauritius considers modernisation of that society as its economy changed. Stanwix and Connell present an unusually detailed consideration of remittances within a MIRAB system.

Aldrich, R. and Connell, J. (1998) *The last colonies*, Cambridge University Press: Cambridge.

Alladin, I. (1993) *Economic miracle in the Indian Ocean: can Mauritius show the way?*, Editions de l'Océan Indien: Rose Hill, Mauritius.

Bertram, I.G. and Watters, R. (1985) 'The MIRAB economy in South Pacific microstates', *Pacific Viewpoint*, 26, pp. 497–519.

Bikoo, S. (1997) 'Mauritius remains India's friend', *Focus Mauritius, A Times of India Country Report*, 12 March 1997.

Encontre, P. (1999) 'UNCTAD's work in favour of Small Island Developing States', *Insula*, 8, 3, pp. 53–7.

Evers, H.-D. (1995) 'The growth of an industrial labour force and the decline of poverty in Indonesia', *Southeast Asian Affairs*, pp. 164–74.

Falklands Island Government (1998) *Report of the Government on the financial year July 1997 to June 1998*, Falklands Island Government: Stanley.

Fedarko, K. (1993) 'Fishing gone', *Time*, 26 April, pp. 66–8.

Financial Action Task Force on Money Laundering (2000) *Review to identify non-co-operative countries or territories: increasing the world effectiveness of anti-money laundering measures*, Organisation for Economic Co-operation and Development: http://www.oecd.org/fatf/pdf/NCCT2000_en.pdf

Firman, T. (1999) 'From global city to city of crisis: Jakarta Metropolitan Region under economic turmoil', *Habitat International*, 23, 4, pp. 447–66.

Grundy, C. and Perry, M. (1996) 'Growth triangles, international economic integration and the Singapore–Indonesian Border Zone', in D. Rumley *et al.* (eds) *Global geopolitical change and the Asia Pacific*, Avebury: Aldershot.

Hampton, M.P. (1996) The *offshore interface: tax havens in the global economy*, Macmillan: Basingstoke.

Hampton, M.P. and Abbott, J.P. (1999) Offshore finance centres and tax havens: the rise of global capital, Macmillan: Basingstoke.

Hampton, M.P. and Christensen, J.E. (1999) '*Treasure Island* revisited. Jersey's offshore finance centre crisis: implications for other small island economies', *Environment and Planning A*, 31, pp. 1619–37.

International Media Corporation (2000) 'World focus: Turks and Caicos special advertising section', insert in *Time*, 22 May 2000.

Isle of Man Bank (n.d.) *A short guide to investing on the Isle of Man*, Isle of Man Bank: Douglas.

Johnson, M. (1979) 'The co-operative movement in the Gaeltacht', *Irish Geography*, 12, pp. 68–81.

Jones, H. (1989a) 'Mauritius; the latest pyjama republic', *Geography*, 74, pp. 268–9.

Jones, H. (1989b) 'Fertility decline in Mauritius: the role of Malthusian population pressure', *Geoforum*, 20, pp. 315–27.

Jones, H. (1993) 'The small island factor in modern fertility decline: findings from Mauritius', in D.G. Lockhart, D. Drakakis-Smith and J. Schembri (eds) *The development process in small island states*, Routledge: London, pp. 161–78.

MacGillivray, D. and Company (1989) *Main price list*, D. MacGillivray: Balivanich, Western Isles.

MacKenzie, D. (1995) 'The cod that disappeared', *New Scientist*, 16 September, pp. 24–9.

Matthews, R. (1978–9) 'The Smallwood legacy: the development of underdevelopment in Newfoundland 1949–1972', *Journal of Canadian Studies*, 13, 4, pp. 89–108.

Mitchell, B. (1994) 'Institutional obstacles to sustainable development in Bali, Indonesia', *Singapore Journal of Tropical Geography*, 15, pp. 145–56.

Norcliffe, G. (1999) 'John Cabot's legacy in Newfoundland: resource depletion and the resource cycle', *Geography*, 84, 2, pp. 97–109.

Ólafsson, B.J. (1998) *Small states in the global system: analysis and illustrations from the case of Iceland*, Ashgate: Aldershot.

Potter, R.B. (1993) 'Basic needs and development in the small island states of the eastern Caribbean', in D.G. Lockhart, D. Drakakis-Smith and J. Schembri (eds) *The development process in small island states*, Routledge: London, pp. 92–116.

Province of Prince Edward Island (1999) *25th Annual Statistical Review*, Department of the Provincial Treasury: Charlottetown.

Richardson, P. (1993) 'Clearing a debt', *Time*, 23 August 1993.

Rowe, F.W. (1985) *The Smallwood era*, McGraw-Hill Ryerson: Toronto.

Royle, S.A. (1995) 'Population and resources in Mauritius: a demographic and economic transition', *Geography Review*, 8, 5, pp. 35–41.

Royle, S.A. (1997) 'Industrialisation in Indonesia: the case of Batam Island', *Singapore Journal of Tropical Geography*, 18, 1, pp. 89–98.

Royle, S.A. (1998) 'Offshore finance and tourism as development strategies: Bermuda and the British West Indies', in D. Barker, S. Lloyd-Evans and D.F.M. McGregor (eds) *Sustainability and development in the Caribbean: geographical perspectives*, University of the West Indies Press: Kingston, Jamaica, pp. 126–47.

Savoie, D. (1999) 'Atlantic Canada: always on the outside looking in', in F.W. Boal and S.A. Royle (eds) *North America: a geographical mosaic*, Arnold: London, pp. 249–57.

Selwyn, P. (1983) 'Mauritius: the Meade Report twenty years after', in R. Cohen (ed.) *African islands and enclaves*, Sage series on African modernisation and development, London: 7, pp. 249–75.

Server, O.B. (1996) 'Corruption: a major problem for urban management. Some evidence from Indonesia', *Habitat International*, 20, pp. 23–41.

Stanwix, C. and Connell, J. (1995) 'To the islands: the remittances of Fijians in Sydney', *Asia and Pacific Migration Journal*, 4, 1, pp. 69–87.

States of Jersey (1993) *The international finance centre*, States Printers: Jersey.

9 Islands of dreams
Tourism, the universal panacea for island problems?

Introduction

Tourism is now the world's largest industry. It has been defined as '1. The practice of travelling for pleasure. 2. The business of providing tours and services for tourists' (*American Heritage Dictionary of English*). This captures well the business aspect; the idea that money can be extracted from those travelling. Much of the tourist industry that impacts on small islands is that subdivision of tourism that is holidaymaking, and the chapter will focus on this. Holiday travel tends to be an activity of people who, on the world scale, are relatively wealthy. Large sections of the developed world's populations now consider it almost an inalienable right to take a couple of weeks away every summer; some might add a winter skiing jaunt to their annual cycle of activities. In the developing world, too, people with money might also travel for pleasure; less wealthy citizens of such nations might sometimes travel but this is more likely to be for work or, perhaps, for religious purposes, given their low disposable incomes. As a country's wealth increases, more of its people will tend to start taking holidays, though not necessarily foreign trips; a large proportion of Americans have always stayed within the USA and relatively few have passports. The Japanese by contrast have become near ubiquitous in overseas holiday areas in recent decades. The summer holidays especially are often taken on islands.

The benefits of tourism to the island situation

Tourists make money in, say, a car factory in Turin or an office in Toronto. Some of that money is then spent on holidays that benefit the economy of Crete or St Vincent or wherever the tourist is staying, for the privilege of the factory or office worker eating, drinking, shopping, sleeping and being entertained away from home. The islander working in tourism is thus supported by the foreign car factory or office and not directly by his or her island's resources. What could be better for the resource-poor places that are small islands? The island itself is simply the place where the transaction occurs. One extreme example of this is the island of Heligoland in the North Sea.

The case of Heligoland

Heligoland is a German island of 160 ha in the North Sea, and the livelihood of its c. 1700 inhabitants depends almost entirely upon tourism. Traditionally its economy was based on fishing, but by the late nineteenth century things had changed. At that time the island belonged to the British and its economic state can be gauged by the annual reports made by its Colonial Governor. By the mid-1870s, such reports were focusing not upon fishing, but upon the early tourist business as Heligoland capitalised upon the fashion for seawater 'cures' (Royle, 1989). The Germans gained Heligoland in 1890, swapping it with the British for Zanzibar, a classic case of island powerlessness (see Chapter 7), and continued to develop it for tourism. Then, in both World Wars, it became a strategic asset. Thus, the British assumed possession at the end of World War II, and handed it back only in 1952. The Germans repopulated it, perhaps initially for reasons of national pride, and built a classic 1950s new town on this sandstone speck in the North Sea. There was obviously a need to re-establish an economy. What better for a place with no resources than to resort to its old industry of tourism? Seawater cures had had their day, but Heligoland was handed another type of tourist activity: duty-free shopping, for it became a duty-free area capitalising upon the day-tripper market. The trippers are principally from Germany and Denmark and the vast majority take a ship for Heligoland from the German mainland or from the nearby holiday island of Sylt. Tourists disembark from jolly boats – the cruise ships cannot tie up – and first usually walk round the island (anticlockwise, people almost always get off boats onto small islands and turn right, favouring the usual dominant side). This is a stroll of an hour or less and they then go to the small town and buy refreshments and goods, before re-embarking, carrying with them items brought to the island solely for the tourists to take away again. Much of the money spent on Heligoland is taken off the island and benefits suppliers and travel companies based elsewhere. This process is called leakage and is a drawback for tourist islands whose environments and societies can be exploited largely to benefit outsiders. However, tourism has become a common activity for small islands, overcoming as it does some of the problems of the lack of domestic resources, and some money and jobs must always stay on the islands.

Goods, services and jobs

Visitors to Heligoland and other holiday areas need goods and services either to consume during their stay or to take away with them; islands and islanders engaged in tourism make efforts to satisfy these needs. There are additional benefits if there are island products involved, say an island beer or other local foodstuffs, crafts or textiles. In such circumstances the tourist proves a convenient buyer, for they travel to the goods and the island producer does not have the cost, bother or risk of having to send the products off the island. Island production is usually small scale and with the inescapable extra costs involved in getting goods to an off-island market, an island may often be unable to compete satisfactorily with mainland products (see Chapter 8). Selling to those who travel to the island is thus an ideal situation, even if some tourists may be unwilling to be educated to local tastes, particularly regarding foodstuffs. Mallorca has a local cuisine and some tourists learn to take pleasure from island delicacies such as *sopa mallorquina* (more of a vegetable stew than soup) or *frito* (fried potatoes, offal, vegetables and herbs). However, with a high concentration of tourists in the mass market, it also has to provide those who wish it with food and drink indistinguishable from that they would eat at home, including burgers and 'tea as mother makes it'.

Tourist demand can distort more than traditional food provision and island farmers might replace their former systems of production with, say, soft fruits or salad crops for tourists if greater profits ensued thereby. Larger scale producers elsewhere may also cash in and, on Mallorca, much of the fresh milk tourists consume comes from Asturias on the Spanish mainland. However, at least in terms of marketing opportunities, if not always in actual delivery, tourism offers on-island agricultural producers an additional market.

Tourists can also provide a market for crafts. If there is a local speciality, tourist demand may even help preserve traditional skills. Thus on the Marshall Islands in the Pacific there is a strong tradition of basketry and hangings, made out of local materials, usually by women. These goods can be made year round in the home and provide a useful source of income even to those on the off-islands rarely visited by tourists, most of whom visit only the capital island of Majuro (Marshall Islands Visitors Authority, 1997). The tourists, too, now form the market for the notable Marshall Islands 'stick charts' which were used as teaching aids to train local seafarers the navigational skills once necessary to sail the Pacific waters. With GPS and other more modern systems, such traditions are no longer needed, but the skill of constructing the charts remains, if only in a bastardised form for tourists. One could denigrate this, of course, as hardly representing a true cultural survival, but without tourist demand – vacationers picking one of these charts from the wall of the airport gift shop – they probably would no longer be made at all, which would surely be worse.

Tourists provide job opportunities for islands in many ways. In fact, on some islands, such as Providenciales in the Turks and Caicos Islands discussed in Chapter 8, on-island job opportunities may overwhelm local labour reserves and thus necessitate in-migration. One employment opportunity is in the construction of tourist accommodation, as well as infrastructural projects, anything from sewers to airport extensions. If the effective population of a holiday island is doubled or trebled in the tourist season, then the infrastructure will have to be adapted to be able to cope with this peak demand. Upgraded water, sewage and transport schemes may well be needed, which might not have been provided for on local demand and such investment is another positive benefit to islanders.

Tourism also provides work directly: hotel workers, restaurant staff, bus drivers, tour guides, taxis, shopworkers, bar staff, etc. Such work may not always be well paid but it might be seen as preferable to work in traditional island occupations such as agriculture and fishing. Its availability may help keep on the island young people who would otherwise have been forced to leave to find work, or who would have chosen to leave to find better opportunities than those the traditional island would have provided. Islands that suffered from out-migration might see their populations restored.

Transportation and cruising

Some familiar island problems are actually less significant than usual regarding tourism: those of peripherality and isolation, for example. True, not all holiday islands can tap into the day-trip or weekend market for foreign tourists; it is not like, say, Belgians driving a few kilometres across to France for a break. With the exception of some from Réunion, people do not go to Mauritius for the weekend. However, with regard to the annual holiday type tourism, the isolation problem is not an issue. Much international tourism involves air travel. It makes little difference operationally to a couple from London if they book to go to the Spanish mainland and their plane lands at Malaga airport, or if they book their holiday on Mallorca and their plane lands at Palma; the distances and, thus, the prices are not dissimilar. All an island needs is direct charter flights from tourist source areas and it can compete on a level playing field with main-land holiday resorts in the same region, only missing out on the land-bound holidaymaker who wishes to drive or go by train.

Some islands without their own airport strive to get one to tap into the tourist market. The state of Malta has three inhabited islands – Malta itself, Gozo and tiny Comino. Gozo is keen to open its own airport. In practical terms this seems hardly necessary. Luqa Airport on Malta has a good range of sched-uled and charter services and anybody wanting to get to Gozo drives to the north of Malta and catches one of the many ferry sailings – it is only a short crossing. There is also a helicopter service from Luqa to Gozo heliport. However, the tourist industry on Gozo is handicapped by a perception that it is difficult to reach and also by the many intervening opportunities on Malta. Why drive a fair way and have to get on a ferry when you are tired after a long flight, or go to the trouble of taking a helicopter, when you can find good accommo-dation in Malta itself near Luqa in Sliema or St Paul's Bay? Pearce (1987) noted that much tourist accommodation tends to cluster by the main urban centre or close to the airport. Thus Gozo does not get many long-stay tourists; rather, most of its holidaymakers take day trips from Malta. If it had its own direct flights it would be easier to establish its own client base who would stay on Gozo for the sun and sand part of their holiday, as well as eat and sleep there, and take their day trips to Malta.

Another important sector of international tourism is cruising, and the journey to the island on the cruise ship is an integral part of the holiday. The ship then disgorges the tourists, credit cards to the fore, when it reaches the island harbour. In Hamilton, Bermuda, the huge cruise liners tie up right at Front

Street and the nearest shops must be about 20 metres from the gangplank (Figure 9.1). It is true that islands are no better off here than mainland coastal resorts, but at least they are no worse off, unless their isolation is extreme. The cruise market is a growing sector of the holiday trade and in regions with many islands, this market offers a welcome opportunity to scale up. Holidays offering calls at a number of different islands in, say, the Mediterranean or the Caribbean mean that each island does not have to market itself individually and economies of scale can be enjoyed, at least regarding marketing. Cruise ship passengers also have the benefit of arriving with their own accommodation, including toilets, and so the island does not have to provide these facilities. To many islands, cruise ships are a welcome component of the tourism industry. In the mid-1990s the Falkland Islands were getting well under 200 land-based holiday-makers per year, but around 6000 arrived on cruises (Royle, 1997). There are even cruises to islands that are barely visited, such as St Helena in the Atlantic. The mighty *Queen Elizabeth II* has visited Jamestown on St Helena: it is a busy day in the tiny capital's pubs, museum, and shops, including the craft shop, and a good deal of money is transmitted to the local economy. This island could not cope with a tourist influx of that size if accommodation was required, but that is handled by the ship. St Helena is not that common a tourist destination for cruising, but Bermuda is, and its marketing is carefully designed to keep its share of this very important trade (Royle, 1998). Bermuda is isolated; cruises there cannot offer the novelty of a different island every day as cruises to the Caribbean do. The Bermudans thus make a virtue from necessity and sell the

Figure 9.1 A cruise ship at Front Street, Hamilton, Bermuda, 1994

long trip over the Atlantic to the tiny archipelago as an integral and enjoyable part of the experience.

The island of dreams and other attractions

In other ways islands are better off than mainland competitors because of their innate romanticism. Where better for a tourist to get away from it all physically as well as spiritually than to go to a separate piece of land? Add in the island of dreams ideas, as discussed in Chapter 1, and islands become attractive destinations. Some travel companies focus on islands as a selling point. If islands are situated on the fringe of a significantly wealthy continent all the better, especially if their location blesses them with warm climates: thus the Mediterranean to Europe, and the Caribbean – that American lake – to North America and the wealthy of Latin America. Australasia and Asia, too, have their holiday islands, from Phuket off Thailand to Tahiti in French Polynesia. There are holiday islands off Africa, such as Mauritius (Wing, 1995; Brown, 1997) and the Seychelles, but here most of the tourists would be long-haul visitors, rather than people from the less developed neighbouring continent.

Butler views the appeal of islands thus:

> irrespective of the exact nature of the appeal, psychological, geographical, economic, climatic, sensual or even convenience, many islands, especially small islands, have become international tourist destinations.
>
> (1993, pp. 71–2)

In the 1970s Air Pacific advertised the 'paradise islands' of the 'romantic South Pacific'. The graphic used depicted all the islands as the same, all uniform high islands, even the atolls of Tonga and the (then) Gilbert and Ellice Islands (now Kiribati and Tuvalu). The smallest in size on the graphic was Papua New Guinea, by a considerable margin the biggest island in reality (Figure 9.2). This was a mental map, based on romantic image rather than geographic realities (Rapoport, 1977). But on such mental maps, on such an unreal depiction of Jamaica as Negril or Haiti as Labadee, both coastal tourism enclaves, islands make their money; or at least money is made on the islands; leakage sees to it that it does not all stay there. And for islands which have no enclaves or romantic image? They have to make the best of what they have. Grenada in 1989 was still struggling to overcome the negative image from its American invasion in 1984:

> it is an uphill struggle, for even if the political issues have faded from popular consciousness, the negative associations survive in the minds of potential visitors.
>
> (*Independent*, 28 October 1989)

Once a war is safely moved into the more distant past, it just becomes something else for islands to exploit. This is common throughout much of the

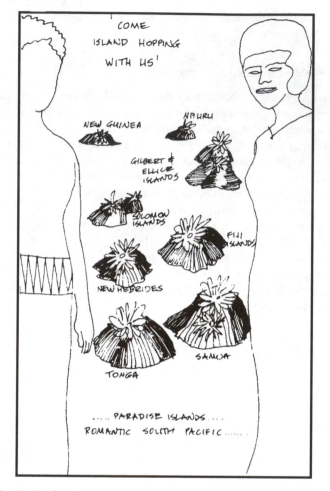

Figure 9.2 An Air Pacific advertisement from the 1970s

Source: Air Pacific

Pacific. Thus the Solomon Islands, scene of near 20,000 deaths at the Battle of Guadalcanal, now markets

> Second World War tourism, vintage Coke bottles to sell to vintage veterans, coral crusted dive-wrecks galore and, of course, megabucks Hollywood films shot on location.
>
> (*The Sunday Times*, 28 February 1999)

Small islands have to market whatever they have: visit Stewart Island, New Zealand's third and most southerly island, if for no other reason than, as you are reminded by the local airline, because it is 'as far south as you can go'.

By contrast, island tourism in regions such as the Mediterranean and Caribbean is very big business indeed. Mallorca is one of the wealthiest parts of

Spain; the Balearic Islands not qualifying for regional aid from the EU; Mallorca's Gross Interior Product per capita (the equivalent of GDP per capita) puts the island on a par with middle-ranking EU nations such as the UK. Bermuda, a British Overseas Territory specialising in tourism and offshore banking, is considerably more wealthy than its colonial motherland; about one-third of its GDP and two-thirds of employment come from tourism (Conlin, 1995; Royle, 1998).

To maximise economies of scale, islands in the usual way of island economics as discussed in Chapter 8, may have to go into tourism in a relatively big way. Liu and Jenkins (1996) show that there is a statistical relationship between population size and the contribution of tourism receipts to GNP. They find that 'the smaller the country [or island], the higher the priority ... given to the development of the tourist industry' (p. 100), often 'as a means of diversification of visible exports' (p. 99), to perhaps provide alternatives to a troubled traditional economy. One example is St Lucia, which has recently diversified its economy from bananas by encouraging tourism (Dann, 1996). In 1997 cruise ship visits increased 37 per cent; in 1998 they increased another 17 per cent and 400,000 people visited. The tourist industry now employs about 12,000, about 17 per cent of the workforce. A similar focus on just this one activity can be seen for many other islands, see for example the case studies in Briguglio, *et al.* (1996) or Lockhart and Drakakis-Smith (1997). Butler (1993) notes that almost all Caribbean states qualify as 'tourist countries' using Bryden's (1973) criteria.

In sum,

> Many countries, especially some small islands are heavily dependent upon international tourism earnings to maintain and increase their levels of income and employment and to obtain foreign currency. Many of these smaller countries would experience great difficulty in developing alternative uses of their resources which would create comparable levels of income and employment.
>
> (Archer, 1996, p. 14)

However, it must be realised that tourism is not without drawbacks in the island situation and to adopt a balanced approach, we will now consider the problems it can engender. See, for example, Lockhart (1997); see also Deda (1999) who considers the double-edged impact of tourism on SIDS (Small Island Developing States).

The drawbacks of tourism in the island situation

There are problems regarding tourism and islands. Even the economic benefits may be somewhat illusory as often much of the money spent on the islands may be repatriated as leakage to large multinational companies who might well dominate the island's industry. The jobs provided are predominantly low skilled and poorly rewarded, often with few prospects. They also often put islanders in a subservient position to the visitors, which can cause resentment. The author was told once with considerable vigour by a community development worker on Prince Edward Island that he did not want the people of his region to have only the opportunity of becoming chambermaids. There are obviously pressures on an island's environment caused by tourism. These are not necessarily relieved by ecotourism or 'alternative tourism' strategies, as Duval notes in his study of St Vincent, as the dispersal of tourists usual in alternative strategies might just 'increase the environmental damage over a larger area' (1998, p. 54). This section will examine the drawbacks of tourism, starting with over-specialisation.

The risks of over-specialisation

The tourist trade is not just large, but it is also varied. Foreign tourists include those people whose travel is motivated by curiosity. These people wish to immerse themselves in an exotic culture – 'wanderlust' tourists is one term used for them. At the other end of the spectrum lies 'sunlust' groups whose aim is to soak up some warmth, and whose interaction with local culture may be limited to paying for familiar goods and services in a different currency. Young adults require a different range of goods and activities compared to children and family groups or the retired. Some tourists are wealthy and require high standards of goods and accommodation; others need cheap accommodation. Those who are providing the facilities might specialise in particular types of holiday experience and aim for market segmentation – a target market strategy (Jefferson and Lickorish, 1988). Alternatively providers can aim to supply attractions for a wide range of ages and interest groups. In the island situation, with restricted scale, it may be difficult to achieve the latter strategy, but the dynamics of the strange beast that is the tourism industry might make a specialist strategy problematical.

Some islands have managed to sustain a target market strategy over the long term. Martha's Vineyard, Massachusetts, USA, an island of about 10,000 residents is one example. 'Top people and no Big Macs', was one description of its high status tourism (*Independent*, 11 September 1996); another was 'half of Hollywood, the cream of New York and *le tout* Washington' (Walker, 1996, p. 4). It was to Martha's Vineyard that John F. Kennedy Junior was flying in 1999 when his plane crashed and he and his party were killed. Kennedy, the son of the US President, had featured in Walker's article as a '1990s standard bearer of the old-money establishment' that continued to patronise Martha's Vineyard (1996, p. 6). Thirty years before the plane crash, another member of this 'old money establishment' family, President Kennedy's brother, Edward Kennedy, was also involved in tragedy here, on the island's Chappaquiddick Bridge where the mysterious death by drowning of Mary-Jo Kopechne probably ruined his own presidential ambitions. High prices keep Martha's Vineyard exclusive, although other tourists can come to observe the celebrities on day trips, their numbers being controlled by a ferryboat licensing system. About 100,000 people are on the island at any one time in the summer. Tourism here is a replacement for the long-gone whaling and fishing industries that were once mainstays of the economy.

Martha's Vineyard has thrived despite tourism being notoriously fickle, for its fads and fashions might see a destination lose its social cachet as tourists move on to a newer, trendier location. The tourist literature has developed a number of stage models which show the dynamic nature of the fashion for tourism destinations (the best known is that of Butler, 1980). Islands losing market share have, on occasion, to freshen their appeal by repackaging and relaunching their tourist product, as with Bali (Connell, 1993). The alternative is stagnation, decline and loss. A tourist island does not want, indeed cannot afford that to happen, and many make strenuous efforts to avoid such an outcome. Bermuda's authorities constantly worry about being overcommitted to tourism, though here there is also offshore finance to spread the risk. Bermuda's Department of Tourism collects and maintains copious records of the tourist industry, including questionnaire surveys of its clients, to assess what is needed to sustain its tourism business. There was a 20 per cent downturn in visitor numbers in the early 1990s as a result of depression and a lessening in all international travel by Americans because of the Gulf War. Bermuda promptly ordered a government inquiry to see what could be done. The principal recommendations related to the re-use of a closed hotel site for a modern facility and the intensification of use of pre-existing resources including trying to extend the season (Government of Bermuda, 1994; Royle, 1998).

The case of Mallorca

Mallorca is an island that has had to relaunch itself in recent years. Tourism here had a hesitant start, as in most European destinations. Writer George Sand in her *Winter in Majorca*, claimed that she and her paramour, Frederic Chopin, were the island's first tourists in the winter of 1838–9 (Sand, 1998). Later, when tourism made its journey from an elite activity towards being more commonplace, Mallorca had some success and in the period between the First and Second World Wars, tourism became important to the island especially through cruising to Palma and Porto Cristo in the south. Mallorca was then regarded as a high quality, even exotic destination, and this was capitalised upon by Adnan Diehl who, in the 1920s, entertained Kings and potentates at his exclusive Hotel Formentor on the Formentor Peninsula, accessible only by boat from Porto Pollensa. He would take rich traders, too, of course, and even advertised his hotel in lights down the side of the Eiffel Tower in Paris. The Spanish Civil War from 1936 and, later, World War II put an end to such fripperies as tourism. By the late 1950s, increasing wealth in Northern Europe meant that some citizens could begin to seek foreign holidays again and transport technology, particularly regarding aircraft, made access to places such as Mallorca possible. At first any foreign travel was exotic and for Britons or Germans to go to the island of Mallorca for their holidays rather than the Isle of Wight or the island of Sylt brought much social cachet. Then came the 1960s, that decade with memorable music but notable, too, for its deleterious impact on the environment, from the dreadful destruction wrought upon urban landscapes to the spoliation of many parts of the world through the unfettered development of mass tourism. Mallorca cashed in. Much of the great sweep of Palma Bay was covered in cheap hotels; patches of tourism sprouted elsewhere around the coast; in came down-market tourism, a place selling to the British fish and chips, British beer and indeed 'tea as mother makes it' (Figure 9.3). Adnan Diehl would presumably have hated it: his hotel remains expensive, but is now accessible by coach. Mallorca made money. Even from people paying relatively little per head, money can be made if there are enough of them. Visitor numbers rose from 100,000 in 1950 to 6.5 m by 1990 and almost 7.5 m in 1998 (Table 9.1).

There was always more to Mallorca than 'tea as mother makes it', and occasionally newspapers would allude to this: 'Unspoilt isle beneath the packaging' was one comment (*The Times*, 3 August 1985). Inland, where few of the mass-market tourists ventured, there remained the characteristic nucleated market towns sitting within fertile agricultural areas and the mountains to the north retained all their majesty. But it seemed that, to most people, Mallorca became associated with just down-market tourism. This is a particularly risky strategy as its success depends on high numbers, on keeping up market share. Mallorca's tarnished reputation put off many of the European bigger spenders who, as transport infrastructure improved, had more choices anyway, and began to go further east in the Mediterranean or to the Caribbean, etc. And then as other holiday areas' marketing strategies continued to develop, even Mallorca's now traditional clientele began to be able to reach more remote places; many people

Figure 9.3 Mass market tourism, Palma Nova, Mallorca, Spain, 1995

Table 9.1 Tourism to the Balearic Islands, 1960–98

Year	Visitor numbers	Revenue (millions of pesetas)
1960	399,831	n/a
1965	1,080,545	n/a
1970	2,274,137	75,907.1
1975	3,435,799	94,699.8
1980	3,550,639	97,813.5
1990	6,068,700	584,966.0
1997	9,242,400	700,000.0
1998	10,142,900	n/a

Source: From information published by the Govern Balear, Conselleria de Turisme.

Notes:
In 1997 tourism to the Balearics produced 60 per cent of the island's GDP.
In 1997 Balearic tourism earnings were about 20 per cent of the Spanish total.
In 1998 Mallorca alone had 7,297,900 visitors entering by air and 143,000 coming by ship.

who would have gone to Mallorca now instead go to Florida. Mallorca was then facing stagnation and had reached 'a crisis point by the mid-1980s' (Bruce and Cantallops, 1996, p. 241). Table 9.1 shows a period of very little growth between 1975 and 1980, for example, for the Balearic Islands. It was a classic case of an island at the latter stages of Butler's (1980) tourism stage model. Tourism had been good to the Mallorcan economy, but the product needed to be relaunched.

> The objective of simply increasing the number of tourists was no longer seen as self-evident and a revised objective of increasing the revenue and employment from tourism . . . was actively canvassed. The Balearics, while continuing with mass tourism, would aim for and find additional up-market segments.
> (Bruce and Cantallops, 1996, p. 248)

A number of acts passed by the local autonomous government placed restrictions on new hotel development, encouraged the modernisation of hotels and aimed to enhance the quality of the tourism experience as well as protecting the environment.

Advertising strategy now stresses the environment and the quality of the tourism product. Calvia, the capital of the district including the classic down-market resort of Palma Nova now hangs banners proclaiming Calvia to provide tourism of quality. Cycling is encouraged, as are nature walks, hikes, golf, and outdoor activities of all sorts. Mallorca strives to cultivate a green image as an environmentally friendly place. That the relaunch has at least changed the reported image is clear: 'Mallorca [was] no longer a byword for cheap and nasty package tourism' by 1999 (*The Sunday Times*, 13 June 1999). This relaunch, as noted, sought to bring in not absolutely more tourists, but a different mix, with a higher proportion from wealthier social groups who spend more. However, Juaneda and Sastre (1999), researchers based at the local university, would caution against treating all segments of the market as equally responsive to the same strategy with their analysis of the different requirements of the German and British segments of the market. To increase tourist numbers would not be a popular strategy domestically; in November 1998 about 60,000 people, one-tenth of the native population, demonstrated in Palma demanding an end to new tourism development. Mallorca, like so many islands, depends now to a dangerous extent upon this one industry. It must and does strive mightily to see that all the eggs in its tourism basket remain intact (Royle, 1996; Bull, 1997).

Tourism's impact upon the insular environment

Another problem relates to the pressures of tourism upon the insular environment. Many islands get many more tourists per year than they have inhabitants. This means that their infrastructures have to be geared up to deal with what in practical terms is a huge population increase. Providing accommodation and recreational facilities is relatively easy. There are often more problems with regard to the impact of the 'population increase' upon natural systems, matters such as sewage and waste disposal and the demands upon the water supply. Mallorca built a new incinerator in the late 1990s but almost immediately it was overwhelmed by the 39,000 tons of garbage generated annually. Tourists tend to like warm islands especially those with little chance of rain during their holiday period. Thus in Europe, the Mediterranean islands with their winter period of precipitation followed by a long hot summer are ideal for tourism. However the environmental drawback is that pressure on water is at its maximum when precipitation is least (Busuttil, *et al.*, 1989). It is not that the tourists necessarily drink much tap water, but they get hot and sweaty and so take regular showers, and, naturally, they use the toilet a lot. Islands have small water catchments; short rivers, perhaps even no surface drainage; they have small lakes, if any – there is just one tiny, if deep, freshwater lake on Crete, for example. They may well have groundwater supplies but again, being in an island situation can cause difficulty, particularly if excessive extraction leads to water table lowering and the problem of salt water infiltration into the aquifer from the fringes of the island. Further, tourism can lead to an increased demand upon water from island farmers if they focus upon products for the tourist trade, such as vegetables and fruit; in hot islands these would certainly require irrigation.

An example of an island under stress over its water supply is Tenerife in the Canaries:

> Tenerife and its resorts, although they have the relevant infrastructure to support the tourism industry, are faced with serious problems – particularly as regards water supply, waste collection and treatment, communications, health care, noise pollution and public safety.
>
> (McNutt and Oreja-Rodriguez, 1996, p. 264)

Tenerife faced water shortage problems for several decades and made plans for extensive action in 1938, 1960 and 1967. But only in 1979, when the island had become a major tourist destination and there was much competition between the tourism and agricultural sectors for water, was the nettle grasped. It took a 3.24 m cubic metre, ten-reservoir scheme, *El plan des balsas del norte de Tenerife*, built from 1980 until the early 1990s, for the problem to be eased somewhat. The cost of $21 m was shared between the autonomous regional government and the Madrid government. The role of the reservoirs is to store rainfall, largely for agricultural use, freeing other water resources for the tourism and urban sectors (Higham, 1992). McNutt and Oreja-Rodriguez support a policy of 'tourist capping' to limit further environmental damage to Tenerife (1996, p. 280).

One of the modern facilities demanded by tourists are golf courses, which, on a hot island, can lead to huge demands on water to keep the greens in good condition. A notice stating in English, Greek and Turkish that nearby playing field facilities were irrigated by water emanating from a sewage works is a memory of the author's first visit to Cyprus in 1986. Environmental pressure groups on Mallorca will probably see to it that the 40 golf courses once planned for the island (*Mallorca Daily Bulletin*, 2 April 1991) will not all be built. In 1995 a huge sign strung across a street in Sa Pobla, one of Mallorca's inland market towns, protested that water is for people, not golf courses (Figure 9.4).

Even tourists' feet can be problematical through the trampling of dunes and other sensitive ecosystems. On the Isles de la Madeleine in Québec, Canada,

Figure 9.4 Protest banner – water for people not golf courses – Sa Pobla, Mallorca, Spain, 1995

most of the islands are interconnected by tombolos. Roads run along these slender and vulnerable links. It is therefore necessary to strive to protect the communications by ensuring the tombolos are not eroded and possibly breached. Thus, for example, on the road between Ile Cap aux Merles and Ile Havre Aubert, bridges link car parks to the actual beaches, stretching over the dunes to protect them from trampling. Environmental problems are common to all tourism areas, of course, but as Chapter 2 explained, islands have particularly sensitive ecosystems.

Islands coming more recently into tourism can learn from the past. One interesting case is the Falkland Islands. We have already seen that cruising is becoming important here. These are basically Antarctic cruises that call in at Stanley on the way. It would be the last port of call, just as it has been for many Antarctic explorers: the island as stepping stone idea again. The passengers can cause temporary overcrowding problems in Stanley, which is a small town of well under 2000 people. However, the real problems are fears that visits elsewhere could harm the islands' environment to the detriment of both the ecology and the islands' future prospects as a tourism destination. Thus tourist parties from the cruise ships are strictly controlled regarding where they can land and they are kept away from sensitive areas where their presence could cause particular harm. Land-based tourists can roam more freely, but their potential impact is kept in check by there being a limit on their numbers of 500 per year. This figure has not yet been approached but is there for the future nonetheless.

The impact of tourism on island society

On Martha's Vineyard, Walker reported on 'old island curmudgeons sporting T-shirts that read "I am not a tourist. I live here and I don't answer questions"' (1996, p. 4). Such dissonance can become a serious problem if an island's traditional economy, along with the social mores and traditions with which it was associated, is transformed by tourism. This can cause resentment: a native Hawaiian group, *Ka Lahui*, does not welcome *haole* (whites) to the islands because of the despoliation tourists have wrought to their environment and traditional society – some remove from the *aloha*-style welcome such tourists have come to expect.

One response to such potential clashes in developing world islands is to isolate the tourists from the islanders by confining tourism to coastal enclaves where the only locals to be seen are wearing uniforms. There is a three-metre wall between the cruise ship port of call of Labadee on Haiti and the rest of the island, for example. There are more than ten all-inclusive resorts on St Lucia which bar non-residents and encourage clients to stay within the resort area. This causes resentment to locals, including those who own other tourist facilities.

> They put nothing into the local economy, import all their own food from the US and their rattan furniture from Malaysia, alienate the islanders who

don't work for them, and – worst of all – provide a sort of artificial holiday experience in which the guests could be anywhere in the world.

(Colin Hunte, a St Lucian holiday villa operator, interviewed in
Independent on Sunday, 1 November 1998)

The author recalls eating at a jerk chicken stall in rural Jamaica when an American couple stepping from a hire car asked for food. They had escaped from their inclusive resort at Negril or Ocho Rios, determined, unlike most, to see something of Jamaica. But they had no local currency; there was no need of such exotica in the resort. They asked the price of the chicken; it was given in (Jamaican) dollars. Would the trader take US dollars? 'No problem,' was the very prompt response. The number of dollars charged remained the same and the trader made a substantial extra profit. The Americans were ripped off, but perhaps people who travel to an island without even knowing what the local currency is get what they deserve. And these were people who were making the effort to see the country.

Most tourists' interaction with island heritage is through a commercialised bastardisation of island traditions: *sega* dances at hotels in Mauritius; fire walking shows in Fiji; aboriginal peoples' dancing displays in the mountains of Taiwan. 'Cultural commodification' and 'staged authenticity' Duval called the equivalent 'traditional' Carib displays on St Vincent (1998, p. 53). Dann (1996) studied the calypso-style West Indian jump up held every Friday night for tourists at Gros Islet on St Lucia. He questioned St Lucians about the tourists' impact upon their culture and found that the results were mixed, with some seeing tourism as providing a useful outlet for its expression. Others were concerned that tourism was unsettling to the local people and their culture.

In an article based on sustainable tourism on Okinawa, Kakazu warned that:

> If we leave the tourism development entirely to sheer market mechanism, the end results will be mass-disruptive tourism which may destroy the very fabric of the local cultural and environmental assets on which islands' tourism industry depends.
>
> (1999, p. 20)

Social problems may be caused by the behaviour of tourists, especially on islands that have become hangouts for downmarket, youthful, 'lager-lout' tourists. Young women may be interfered with; there could be crime; maybe drugs; there will certainly be drunken rowdiness and damage. Islands often enjoyed a conservative society; to be exposed to unleashed northern European holidaymakers with all their inhibitions cast aside is a fearsome prospect: 'I hate them, I really hate them. The English behave like pigs. They respect nothing. They know nothing' (an Ibiza hotel receptionist quoted in the *Guardian*, 1 September 1998). 'Holiday island where British means brutish' was the headline on an article which stated that Ibiza had 'prostituted its culture and environment, [but was] still entitled to be shocked by the brutishness of the British clients' (*Independent*, 3 August 1998). Both German and British television showed programmes in 1998 'celebrating' the activities of their young people on Ibiza. Such were the pressures on him by tourists' bad behaviour that in August 1998, Michael Birkett, the British Vice-Consul on Ibiza 'quit in disgust at his country-men's depravity' (*Guardian*, 1 September 1998). To be balanced, it should be noted that the newspaper also quoted a *Diario de Ibiza* reporter saying that only a 'tiny minority' of the English caused problems.

The effect of tourism on traditional island economies

It was noted above that island agriculture might be helped by tourism in that new markets are developed. However, there are circumstances in which tourism may just provide competition for labour, in which case an island's agriculture or other traditional economic activities may be harmed. Islanders themselves may give up agriculture. Agriculture on Mallorca, and also on Guernsey, is now at least partially dependent upon foreign workers to supply the labour force. On the Irish island of Inishmore, the economy is now driven by tourism, no longer by primary production, and the old ways are being lost. Field systems took immense effort to bring into being here, including the backbreaking process of 'landmaking' (as explained in Chapter 4). Now these fields are neglected, for agriculture is little more than a pastime. In one of the many places celebrating what is really now no longer an extant cultural heritage, the Dún Árann Heritage Centre, exhibit number 7 in 1997 was actually a patch of potatoes, grown to demonstrate the crop that used to be important here. Now, with modern tourism, the island is obviously better off, and, indeed, of all the Irish islands it is Inishmore that is probably doing best in economic terms and with regard to retaining its population. However, people from more traditional islands such as Inishturk wince at the way in which tourism has transformed Inishmore's society and traditions, despite the obvious benefits from tourism's economic opportunities. It is romantic nonsense, and outsiders' romantic nonsense at that, for it to be suggested that islands like Inishmore should not change, should remain true to their traditions. The Irish coast is littered with islands where doubtless traditional society was maintained right up until the moment depopulation ended all society on the island as its traditional ways could not support the population to standards readily available elsewhere. Islands like Inishmore may well have to adapt or die, and it is islands like Inishturk where the population is still basically maintained by agriculture and fishing that are now unusual. However, the cost of Inishmore's selling out to tourism has been that the island has become a museum of itself. Locals gain their living from driving minibuses and horse-drawn carriages to show tourists who do not speak their native Irish or, in many cases, their second language of English, how their parents or grandparents used to live.

The case of Kulusuk Island

Let us conclude this chapter with a consideration of one of the smallest and most remote insular communities to be affected on a daily basis by tourism.

This is Kulusuk Island, which lies off the shore of eastern Greenland. Many of Greenland's approximately 80 settlements are actually not on the mainland of Greenland itself. The usual benefits of being on the larger landmass with regard to connectivity simply do not apply to Greenland. No two Greenland settlements are connected to each other by road. The vast majority of the mainland is completely uninhabitable and, except by specialist expeditions, off-limits to travellers as it is covered by ice. Only in the more favoured – and green – south is there any agriculture. The rest of the settlements, outside the service and administrative work in the capital city of Nuuk and also some military areas, are largely dependent on marine resources. Choice of settlement site under such circumstances is predicated without much reference to the land, only to the land's access to the sea. Thus islands are often chosen. In fact the word *Nuuk* means peninsular, it is as if its location on the said peninsula of the mainland is a fact worth recording. In eastern Greenland, apart from Ittoqqortoormiit, a fairly recent settlement to the north of the others, there are just six villages in the southeast, with about 1400 Inuit inhabitants, and one town, Angmagssalik, with about the same number.

Kulusuk Island has about 350–400 people, all now living in one village, Kulusuk itself, sometimes called Kap Dan. There was another village on the island but it has been abandoned. Kulusuk has eastern Greenland's only airstrip; a gravel runway originally built by the now-departed Americans to service an early warning station based on the island in the days of the Cold War. This airstrip has flights to Nuuk and to Iceland, as well as helicopter flights to the other settlements in the area. Kulusuk is thus the access point for eastern Greenland. As such, the island receives the bulk of the region's tourism. This is mainly in the form of day trips from Reykjavik.

Planeloads of rich tourists from or visiting Iceland descend on Kulusuk every summer, six days per week. There is a 34-bedroom hotel close to the airport, but most tourists are day-trippers who follow a standard programme. This consists of a guided walk along a dirt track from the airstrip to the village; viewing a performance of a traditional Greenlandic drum dance; perhaps also a kayaking demonstration. The tourists then walk or a take a boat to the hotel for lunch and afterwards return to Reykjavik which has, over the preceding few hours, come to seem to be a lot less peripheral and provincial.

Tourism brings money and work into the Kulusuk economy. In the village the one shop is staffed by locals, so they benefit from the tourist trade. Similarly, the Inuit women who await the tourists in the village to sell them crafts benefit directly, with no middlemen to take a share of their profit. Some men sell bone carvings and benefit as well. The men providing the boat rides get paid, as does the drum dancer. However, not all the money spent remains with the locals; most of the airport jobs are carried out by people of Viking (i.e. Danish or Icelandic) appearance, not by Inuits, and the hotel staff are mainly Danes or Icelanders. The buffet lunch is standard Danish fare, the food imported.

Inevitably, there is potential for the tourism to affect Kulusuk's society and culture. However the airline's Icelandic guides inform trippers that the prices the locals first state are fixed, so there is no humiliating haggling and the trans-actions appear free from aggression on either side. The drum dancer certainly bastardises his art; it is viewed out of context, without the traditional symbol-ism, but he seems to enjoy his performance. Other adults are more reserved, but they, too, seem unperturbed by the foreign presence. The children enjoy the tourists. The guide can be observed doing his job with his hand firmly clutched by an Inuit toddler to whom he is obviously a familiar and trusted friend. Chil-dren constantly offer themselves for photographs (Figure 9.5). This is a game, there is no expectation of reward (the huskies, in contrast, tend to be less co-operative in their posing). The traditional economic life of the village continues and there is much evidence of the locals' considerable exploitation of seals. This is to the distaste of some visitors who, presumably, have not considered how Kulusuk people have to make their living.

There must have been changes wrought by tourism to Kulusuk Island. There are some jobs; there is certainly more money in the village as a result. There has been some care taken to minimise tourism impact such as the no haggling rule and the location of the hotel at the airport, out of sight of the village. On the other hand traditionalists might feel that the drum dance should not have become a tourist spectacle and the daily parade of rich foreigners, some with cameras the value of several seals, must also have an unsettling effect. The village has several uninhabited houses, presumably through the out-migration of those unhappy at the prospect of facing the twenty-first century catching seals. It might also be pointed out that the better jobs in the hotel and airport were not held by Inuits. However, for the present, Kulusuk remains a Green-landic sealing island, with tourism as an additional, rather than replacement activity. Other small islands have been more adversely affected by tourism. Perhaps Kulusuk islanders are fortunate that their tourism is controlled from Iceland, whose people are traditionally concerned with welfarism and commun-ity support.

Conclusion

Tourism everywhere is a doubled-edged sword, which can harm the resort area as well as supporting it. In the particular circumstances of islands where the environment and society may both be fragile, it can be a dangerous activity indeed to take on. However, given the limited resources of the typical small island and the unique characteristic of tourism that sees money made elsewhere brought in to spend, tourism is seen as a universal panacea for islands in the modern world. So, a flyer encourages us to haste to Denman and Hornby Islands, the self-declared 'undiscovered islands' of Western Canada's Strait of Georgia. However, what of islanders on Denman and Hornby who wish their homes to remain 'undiscovered'? Another universality in this modern world, the media, will see to it that they get discovered anyway. There are, or at least were, still Greek island 'hideaways if you know where to look', said an article uncover-

Figure 9.5 Inuit children, Kulusuk Island, Greenland, 1999

ing Lesvos, Halki, Leros, Lipsi, Astipaliaia, Pserimos, Nissiros, Lefkas and Lymnos (*Independent*, 28 February 1987). Further, 'in a world where few tourist-free zones remain, there are still deserted parts of the Philippines to explore', said another, encouraging readers to trip over each other on islands off Palawan (*Independent on Sunday*, 1 November, 1998). Another article was at least honest about the process: 'tourism has discovered Cuba, but only just, so visit the island now while the scenery is still unspoilt and the people friendly' (*Independent*, 29 October 1994). Not a hint that if you did not visit at all, the scenery and the people might stay that way. In another article, 'Stan Abbott savours the savage splendour of the Arctic island of Spitzbergen and explains why you should not go and see it for yourself'. The article ended with the usual box detailing travel options to get there (*Independent*, 7 November 1992). However, islanders have to make a living, even friendly islanders cannot live on scenery. Thus we conclude with an Ibiza night club bouncer contemplating two 'bare chested young men gleaming with vomit, "They get sick, so what? As long as they're not violent it's OK. We need these people, without them we've no work. What then?"' (*Guardian*, 1 September 1998).

References

Key readings

A number of recent edited volumes on tourism and islands cover a wide range of theoretical and practical issues as well as presenting a good number of case studies. These are the two books edited by teams led by Lino Briguglio, and others by Douglas Lockhart and David Drakakis-Smith, and by Michael Conlin with Tom Baum.

Archer, B. (1996) 'Sustainable tourism: an economist's viewpoint', in L. Briguglio, B. Archer, J. Jafari and G. Wall (eds) *Sustainable tourism in islands and small states: issues and policies*, Pinter: London, pp. 6–17.

Briguglio, L., Butler, R., Harrison, D. and Leal Filho, W. (eds) (1996) *Sustainable tourism in islands and small states: case studies*, Pinter: London.

Brown, G.P. (1997) 'Tourism in the Indian Ocean: a case study of Mauritius', in D.G. Lockhart and D. Drakakis-Smith (eds) *Island tourism: trends and prospects*, Pinter: London, pp. 229–48.

Bruce, D. and Serra Cantallops, A. (1996) 'The walled town of Alcúdia as a focus for alternative tourism on Mallorca', in L. Briguglio, R. Butler, D. Harrison and W. Leal Filho (eds) *Sustainable tourism in islands and small states: case studies*, Pinter: London, pp. 241–61.

Bryden, J.M. (1973) *Tourism and development*, Cambridge University Press: Cambridge.

Bull, P. (1997) 'Mass tourism in the Balearic Islands: an example of concentrated dependence', in D.G. Lockhart and D. Drakakis-Smith (eds) *Island tourism: trends and prospects*, Pinter: London, pp. 137–51.

Busuttil, S., Villain-Gandossi, C., Richez, G. and Sivignon, M. (1989) *Water resources and tourism on the Mediterranean islands*, European Coordination Centre for Research and Documentation in Social Sciences and the Foundation for International Studies at the University of Malta: Valletta.

Butler, R.W. (1980) 'The concept of a tourist area cycle of evolution: implications for the management of resources', *Canadian Geographer*, 24, 2, pp. 5–12.

Butler, R.W. (1993) 'Tourism development in small islands: past influences and future directions', in D.G. Lockhart, D. Drakakis-Smith and J. Schembri (eds) *The development process in small island states*, Routledge: London, pp. 71–91.

Conlin, M. (1995), 'Rejuvenation planning for island tourism: the Bermuda example', in M.V. Conlin and T. Baum (eds) *Island tourism: management, principles and practice*, Wiley: Chichester, pp. 181–202.

Connell, J. (1993) 'Bali revisited: death, rejuvenation, and the tourist cycle', *Environment and Planning D: Society and Space*, 11, pp. 641–61.

Dann, G.M.S. (1996) 'Socio-cultural issues in St Lucia's tourism', in L. Briguglio, R. Butler, D. Harrison and W. Leal Filho (eds) *Sustainable tourism in islands and small states: case studies*, Pinter: London, pp. 103–21.

Deda, P. (1999) 'Sustainable tourism in Small Island Developing States', *Insula*, 8, 3, pp. 40–4.

Duval, D.T. (1998) 'Alternative tourism on St Vincent', *Caribbean Geography*, 9, 1, pp. 44–57.

Government of Bermuda (1994) *Report of the Premier's Task Force on Employment*, Government of Bermuda: Hamilton.

Higham, S. (1992) 'Last resort', *World Water and Environmental Engineer*, September 1992, pp. 17–18.

Jefferson, A. and Lickorish, L. (1988) *Marketing tourism: a practical guide*, Longman: Harlow.

Juaneda, C. and Sastre, F. (1999) 'Balearic tourism: a case study in demographic segmentation', *Tourism Management*, 20, pp. 549–52.

Kakazu, H. (1999) 'Sustainable tourism development in small islands, with particular emphasis on Okinawa', *Insula*, 8, 3, pp. 15–20.

Liu, Z.-H. and Jenkins, C.L. (1996) 'Country size and tourism development: a cross-nation analysis', in L. Briguglio, B. Archer, J. Jafari and G. Wall (eds) *Sustainable tourism in islands and small states: issues and policies*, Pinter: London, pp. 90–117.

Lockhart, D.G. (1997) 'Islands and tourism: an overview', in D.G. Lockhart and D. Drakakis-Smith (eds) *Island tourism: trends and prospects*, Pinter: London, pp. 3–21.

Lockhart, D.G. and Drakakis-Smith, D. (eds) (1997) *Island tourism: trends and prospects*, Pinter: London.

Marshall Islands Visitors Authority (1997) *Handcraft of the Marshall Islands*, Forum Secretariat: Suva.

McNutt, P.A. and Oreja-Rodriguez, J.R. (1996) 'Economic strategies for sustainable tourism in islands: the case of Tenerife', in L. Briguglio, R. Butler, D. Harrison and W. Leal Filho (eds) *Sustainable tourism in islands and small states: Case studies*, Pinter: London, pp. 262–80.

Pearce, D. (1987) *Tourism today: a geographical analysis*, Longman: New York.

Rapoport, A. (1977) *Human aspects of urban form*, Pergamon: Oxford.

Royle, S.A. (1989) 'A human geography of islands', *Geography*, 74, pp. 106–16

Royle, S.A. (1996) 'Mallorca: the changing nature of tourism', *Geography Review*, 9, 3, pp. 2–6.

Royle, S.A. (1997) 'Tourism to the South Atlantic islands', in D.G. Lockhart and D. Drakakis-Smith (eds) *Island tourism: trends and prospects*, Pinter: London, pp. 323–44.

Royle, S.A. (1998) 'Offshore finance and tourism as development strategies: Bermuda and the British West Indies', in D. Barker, S. Lloyd-Evans and D.F.M. McGregor (eds) *Sustainability and development in the Caribbean: geographical perspectives*, University of the West Indies Press: Kingston, pp. 126–47.

Sand, G. (1998) *Winter in Mallorca*, La Foradada: Palma (first published in French in 1855).

Walker, M. (1996) 'Millions like us', *Observer Life*, 4 August 1996, pp. 4–6.

Wing, P. (1995) 'Tourism development in the South Indian Ocean: the case of Mauritius', in M.V. Conlin and T. Baum (eds) *Island tourism: management, principles and practice*, Wiley: Chichester, pp. 229–37.

10 Conclusion

St Helena wins again

Introduction

In this concluding chapter there is a need to wrap things up. There would seem to be two choices of how this could be done: a reiteration of the themes in summary form or a consideration of them in operation, this time together, by looking in detail at one island where they have all had an effect. This latter strategy is the one adopted. This choice has the benefit also of allowing us to consider one island in detail, a sustained piece that has not been possible to present in the other chapters where the case studies were partial, chosen to illustrate just one particular aspect of insularity – scale, isolation, resource availability, etc.

It is the author's contention that every island is impacted in some way by the range of insular constraints explained in the preceding chapters, but differing in degree depending upon local circumstances. This chapter has to consider one small island where every aspect of insularity looms large. The island chosen, the 'ultimate island', is St Helena in the South Atlantic (Figure 10.1). This is not to claim St Helena as a 'typical' island; there is probably not such a thing anyway. Rather, St Helena is an 'extreme' island, a place where insularity massively affects every aspect of the history, environment, society and economy.

St Helena – geology and ecosystems

St Helena is a product of the mid-Atlantic ridge and owes its origin to the outpouring of igneous material from between the diverging African and Latin American plates. It is thus volcanic, but, unlike Ascension, Tristan da Cunha or Iceland, which sit astride the ridge and are active still, St Helena is now just a relict of such vulcanicity. The island has been carried along by tectonic movement and lies some way from the ridge on the African side. Its landscape, given its geology, is rugged, with steep and rather forbidding coastal margins giving way to a plateau-like interior with much surface at around 400 m, though there are mountains, culminating in Diana's Peak at 820 m. St Helena, at 16 degrees south, is tropical, but its climate is ameliorated by altitude and by the moderating effect of the surrounding ocean. Temperatures and precipitation are height related with the dryer, cooler coastal margins, where precipitation can be

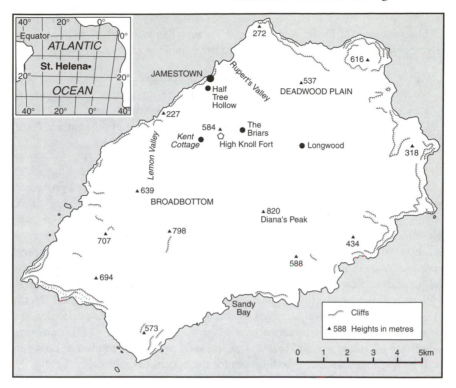

Figure 10.1 St Helena

around 150 mm, giving way to a wetter inland, where orographic effects come into operation and precipitation can exceed 1200 mm. The coast has little vegetation and is not productive. Where the relief flattens out, at around 400 m, there is sufficient precipitation and depth of soil to allow for cultivation, and the cultivable area of St Helena corresponds closely with that area enclosed by the 400 m contour. Thus on an island already small at 121 km², it is only a much smaller area that can be put into production, a factor of some importance in the human use of this island.

St Helena is very isolated. Its nearest land is Ascension Island, 1130 km northwest; its nearest continent is the coast of Angola 1840 km east; Brazil in Latin America is more distant at around 2880 km. This has given St Helena an island-type biogeography with a restricted flora and fauna (for example, frogs, snakes and lizards are absent, though there are geckoes) and it has some endemic species. St Helena lacks anything endemic of the splendour of the dodo and is too isolated to have developed the biological interest of, say, the Galapagos. However, it has species of significance, in particular *Aegialitis sanctae-helenae*, the endemic wirebird, often to be seen, looking rather ungainly on what appear to be over-long, spindly, red legs. There are also nine endemic tree types, including three cabbage trees: the He-Cabbage Tree (*Senecio leucadendron*); the She-Cabbage Tree (*Senecio prenanthiflora*); and the Black Cabbage Tree (*Melanodendron integrifolium*) (Brown, 1981). Many species have been introduced, by no means all to the benefit of the island, thus the:

> mynah, *Gracula religiosa*, from India. Introduced in 1829 to keep the cattle free of ticks. The good they do to the cattle is more than countered by the damage they do to the fruit . . . All too common.
>
> (Gosse, 1938, p. 425; see also Ashmole and Ashmole, 2000)

Humankind was a late addition to St Helena's list of resident species but has had a devastating effect on the island's environment, an example of the fragility of island ecosystems. Human influence was baleful from the outset. The Portuguese, who discovered the island in 1502, stocked it with goats, to be hunted as a food supply by passing ships' crews. The environment, especially the flora, was irreparably affected and feral goats continued to impact the island's natural vegetation until they were finally controlled nearly 400 years later in the 1980s. By that time, centuries of cultivation had further affected the natural ecosystems and the botanical gardens at Kew in London have mounted rescue expeditions to the island. St Helena is threatened neither by hurricanes, nor, at over 800 m, particularly by the effects of global warming; but even so, the fragility of the insular environment is clearly seen. Further, the threats may not be over since Deadwood Plain, the proposed site for the airport St Helena so desperately needs if its moribund economy is to be rescued, is a major breeding area for the endemic wirebird.

Population and economic history: an island story

There are no aboriginal St Helenians; the island was uninhabited, and, given its extraordinary isolation, presumably unseen by humans before being discovered by the vessel of the Portuguese admiral, João da Nova Castella, in 1502. The Portuguese used it as a sanctuary and also as a supply station for vessels returning from India (a process known as 'revictualling'). The island's first resident jumped ship at the island in 1513 on one of those voyages. He was Fernando Lopez, who had been a deserter from a conflict in Goa between Portugal and Indians. As punishment he had had his nose, ears, right hand and left thumb removed and did not fancy what else might be done to him when he was returned to Portugal. Lopez survived for ten years before being found and returned to Lisbon as a celebrity. He had an audience with the Pope at which he expressed his desire to return to St Helena. He was so returned and survived until 1545, alone with his poultry.

Today's Saints (as the people of St Helena are known) descend from people brought to the island somewhat later (Table 10.1). They are a mixed population, a rather unusual true melting pot: 'each new baby is a surprise re-jigging of the genes' (Bain, 1993, p. 31). The genes are those from European settlers and others taken to St Helena from the seventeenth century by the British East India Company; Africans brought as slaves or, later escaping from slavery; Indians and Chinese brought in as labourers; passing soldiers and sailors and any additions to the island's people from its use as Napoleon's prison or as a prisoner-of-war camp for the Boers during the South African War. Much of the initial migration related to the island's use as a revictualling station for the East India Company. That an entire place could be run by a company is a product of island scale; that the company had then to bring in the workers as there were no native people is a factor relating to island isolation. A fairly large labour force was needed as ships could not, and still cannot, tie up at St Helena but have to anchor offshore with passengers and goods being ferried to and fro in small boats. In the Company's time these boats were rowed and there was a need for hundreds of workers on the island. Here, then, was an island whose economic function was predicated upon its strategic position, a not uncommon situation.

St Helena's economic history remained quintessentially insular. It had a period of prosperity from 1815 to 1821 when there was a lot of spending on the island due to its use as a prison for Napoleon (Figure 10.2). After his death, ship revictualling became again the prime occupation with sometimes around 1000 ship visits per year. However, technological change brought an end to that. Better food storage facilities on ships and the replacement of wooden sailing ships by more robust iron and steel ships with engines meant that revictualling stops became less necessary. In any event the opening of the Suez Canal in 1869 took away most of the shipping from the South Atlantic as ships bound for Europe from the east no longer needed to round the Cape of Good Hope (Table 10.2).

St Helena was also of strategic value with relation to Britain's suppression of

Table 10.1 St Helena population totals, 1683–1998

Year	Population
1502	Discovery
1683	500
1723	1110
1814	3507
1851	6914
1881	5059
1891	4116
1901	9850
1911	3520
1921	3747
1931	3995
1946	4748
1956	4642
1966	4649
1976	5147
1987	5644
1991	5399
1994	5270
1998	4900

Source: St Helena Statistical Yearbooks.

Note: The inflated total of 1901 reflects the island's use as a prisoner-of-war camp for Boers captured in the South African War.

Figure 10.2 Longwood House, residence of Napoleon Bonaparte, St Helena, 1990

the slave trade between 1839 and 1874. Africans liberated from slavers' ships were on occasion taken to St Helena and accommodated in Rupert's Valley until they could be returned. With the exception of another brief period from 1900–2 when the island was again brought into service for British military purposes, after the ending of its uses for strategic purposes the economy of St Helena became moribund. Conditions for the Saints were rather poor and it took the efforts of a dedicated Colonial Surgeon, Wilberforce Arnold, at the start of the twentieth century to improve public health matters. This earned him the Saints' undying admiration and a prominent monument in the middle of Jamestown inscribed to 'the best friend St Helena ever had' (Royle and Cross, 1995; 1996). Arnold was conscious of the need for better employment conditions as a tool to improve health and welfare matters. He supported the innovation introduced by Governor Henry Gallwey to re-introduce the cultivation and utilisation of New Zealand flax, *phormium tenax*, to the island. The flax was used largely to make rope and string in mills on the island and St Helena's cultivable area became completely dominated by its production (Northcliffe, 1969). For example, during World War I the

> Saint Helenians grew reckless over the planting and gave up growing anything else. They made a lot of money, but most of it had to be spent buying, at fancy prices, things to eat which before the flax boom they had grown for themselves.
>
> (Gosse, 1938, p. 349)

Table 10.2 Shipping visits to St Helena by type, 1752–1995

Year	Ship visits	Warships	Merchant ships			Whalers	Captured slave ships	Others
			Sail	Steam	RMS			
1752	30							
1774	39							
1793	133							
1802	169							
1815	186							
1823	193							
1830	367							
1834	475							
1840	737							
1845	475							
1850	1008	27	916	0	0	27	38	0
1855	1100							
1860	1044	23	938	14	0	64	5	0
1865	850	23	753	38	0	35	1	0
1870	807	24	672	45	0	66	0	0
1875	604	25	507	41	0	31	0	0
1880	563	26	450	41	0	46	0	0
1885	450	25	363	40	0	22	0	0
1890	211	12	157	26	0	16	0	0
1900	171	4	96	71	0	0	0	0
1910	51	7	8	33	0	3	0	0
1920	37	5	2	30	0	0	0	0
1930	38	1	1	36	0	0	0	0
1940	59	25	0	34	0	0	0	0
1950	26	1	0	25	0	0	0	0
1960	39	1	0	29	0	0	0	9
1970	73	11	0	39	0	0	0	23
1980	26	0	0	2	14	0	0	10
1990	27	0	0	0	15	0	0	12
1994	39	0	0	0	21	0	0	18
1995	30	0	0	0	18	0	0	12

Source: St Helena Statistical Yearbooks.

Notes:
1. Blanks mean there is no information.
2. Yachts are not included.
3. RMS stands for the government-owned Royal Mail Ship, which replaced the liners en route to South Africa which used to provide the island's shipping service.
4. Other ships include tankers, etc.

Flax production became a classic insular 'all the eggs in one basket' scenario. The eggs stayed intact in the almost 60 years that flax was the mainstay of the economy and there were good times, particularly during wars; armies need rope and string it seems. However, in 1966 the then predominant buyer of St Helena string, the British Post Office, decided to use nylon twine and rubber bands to bundle its letters and the market for St Helena's only product of note collapsed – all the eggs were broken.

The present economic situation: an Atlantic MIRAB

Since the collapse of demand for flax, St Helena has not really had a productive economy. A recent visitor to St Helena, writer Harry Ritchie, in a hard-hitting piece condemns the British government for its ineffectual direction of the economy. The Overseas Development Administration, he claimed, 'had mucked around for years and years and years and come up with absolutely bugger all' (1997, p. 227). (The British Overseas Territories are now under the direction of the Foreign and Commonwealth Office and the Department for International Development.) Basically, St Helena cannot support itself, in the 1990s imports were running at several dozen times the value of exports. Ritchie has it thus: 'six thousand people live on St Helena [4900 residents at the 1998 census] and, for the life of me, I could not see how' (1997, p. 223).

There have been projects: coffee and honey are exported, although fish (principally tuna) – frozen, dried and canned – remains the principal export. The coffee is a branded product, sold through The Island of St Helena Coffee Company (http://www.st-helena-coffee.sh/main.html). There is some tourism – the *Queen Elizabeth II* has called, as have other cruise ships – and there is encouragement for local business development and outside investment. However, commerce is dominated by a single company, Solomons, a product of the scaling up needed in an island situation. This company is presently 84 per cent government owned, although that proportion may fall. The domination of commerce by this government company, which is involved in food and general retailing, farming, shipping and insurance, does not help the development of self-employment or of small and medium enterprises in the commercial sector.

The St Helena government has identified the attractions for outside investors:

- strong government support;
- free access of products to the European Union;
- a growing track record of high quality products for export;
- tax concessions for investors;
- low political risks;
- no exchange controls;
- an educated workforce;
- outstanding quality of life (Chief Secretary, 1998).

Perhaps outside investors will be attracted; 16 approaches were made to the St Helena Development Agency by potential investors between 1994 and 1996. There are hopes that coffee and honey exports will expand. Fishing is being reorganised with the input of Argos, a private outside company. There will be privatisation of some government holdings. Electronic business could provide some work, although presently charges for access to the Internet are high. On-island brewing is to be revived. Tourism is to be encouraged, as is import substitution (St Helena Strategic Review, 1996). There are plans, hopes and schemes, as there have always been, despite Ritchie's criticism recorded above. But without an airport to at least reduce their impact, the geographical, insular constraints of St

Helena rule against the island being able to develop a self-sustaining economy. So, St Helena has to be supported from outside. Some of this support is from remittances from Saints who work off the island. In 1999 most, about 600, were on Ascension Island, a St Helena Dependency where Saints form much of the civilian workforce for that island's military and commercial users. Around 400 work on the Falklands where there is a labour shortage; about 300 in the UK. There are also around 90 Saints who have jobs on the government ship, the *R.M.S. St Helena*. So, despite the workings of the *British Nationality Act* with the difficulties it caused regarding migration to the metropole (see below), around 25 per cent of St Helena's workforce have been able to get to work off the island. They contribute between £2 m and £3 m per annum to the St Helena economy. Thus, we have the MI(gration) and R(emittances) of MIRAB (see Chapter 8).

St Helena has to be massively supported by British aid, although the amounts granted have tended to decline in recent years. It is the only Overseas Territory still to receive budgetary aid used to balance the local budget; as well as receiving technical co-operation funds, development aid and a subsidy to support the ship. Aid totals were between £5.5 m and £9.7 m per annum in the 1990s. About 30 per cent of the St Helena government's annual recurrent budget is funded by the UK, as is almost all capital investment. This is the A(id) in MIRAB (Table 10.3).

A survey carried out in 1990 (Royle, 1992) found that there was little entrepreneurialism within the island's young people. Almost all expected to have to

Table 10.3 UK aid to St Helena, 1979–95

Year	Budgetary aid	Shipping subsidy	Development aid	Technical co-operation	Total (£000)
1979–80	1299	1540	510	407	3756
1980–1	1631	1300	396	536	3863
1981–2	1835	1737	776	577	4925
1982–3	2392	1710	855	635	5592
1983–4	2722	2322	1722	566	7332
1984–5	3229	2185	1432	901	7747
1985–6	3664	2064	2836	1467	10,031
1986–7	3421	1484	3351	1602	9858
1987–8	2240	1303	2918	1547	8008
1988–9	3543	3543	1338	1552	2265
1989–90	3904	1685	2191	1986	9766
1990–1	3477	2760	1407	2008	9653
1991–2	3424	920	1398	2187	7929
1992–3	3526	1589	1614	1930	8659
1993–4	3496	1071	1051	2364	7982
1994–5	3225	965	1446	2364	7531

Source: St Helena Statistical Yearbooks.

Note: in addition, £31,744,000 was spent between 1987 and 1991 to provide a new government ship, *R.M.S. St Helena*. From 1997–8 a 3-year Country Policy Plan operated providing £26 m aid in total including £3.2 m p.a. budgetary aid and c. £1.3 m p.a. shipping subsidy.

work for the government, the overwhelmingly predominant employer, with about 84 per cent of the workforce being government employees. This is quite usual for small dependent islands where such an employment pattern 'creates or maintains jobs and activities that have value in sustaining households rather than contributing to economic development' (Aldrich and Connell, 1998, p. 91). The government is also involved in matters that, in larger economies, would be the preserve of the private sector. These include power and water supplies, banking (there is only the Government Savings Bank), commerce (through its ownership of Solomons) and the media (though not telecommunications and television which are provided by Cable & Wireless). This is the B(ureaucracy) in MIRAB and completes the acronym. St Helena is a classic case of a Pacific-style MIRAB economy, but situated in the Atlantic Ocean.

Powerlessness

Lopez apart, and later occasional sick sailors put ashore there for fear they would not survive the voyage up to Portugal, the Portuguese had not permanently occupied St Helena. Thus, when other European nations independently stumbled across the island it was regarded as open territory and other countries began to use it. There were even the occasional skirmishes between them as with the Spanish and Dutch in 1601. The Dutch claimed St Helena in 1633 but also probably did not occupy it permanently and this allowed the British East India Company to assume control in 1659, with the aim of using the island as a revictualling base to service their ships returning from India. 'St Helena became the English (sic) Company's only safe haven, watering place and larder on the direct route between India and England' (Keay, 1991, p. 250; see also Schulenburg, 2001). There was contestation still with the Dutch, and the British were expelled from St Helena on New Year's Day 1673, a typical case of a powerless island being unable to resist an invasion. As with the Falklands Campaign 309 years later in the same ocean, the British mounted a counter-invasion, the forces on the island were unable to resist, and within six months the British had resumed control and occupation (Gosse, 1938). British control has since remained unchallenged, although the East India Company went to considerable expense to build fortifications in case of any further colonial rivalry. Control passed from the Company to the British Crown in 1834.

Powerlessness means more than just an island being unable to resist military threats. In St Helena, the lack of influence of island inhabitants and the lack of consideration given to such people by authorities concerned, perhaps, with more significant matters continues to be shown with regard to a recent situation, not unrelated to events of 1673. After the reconquest of the island, King Charles II issued a Royal Charter on St Helena, one of its provisions being that:

> Wee do for us, our heirs and successors declare ... that all and every the
> persons being our subjects which do or shall inhabit the said port or island,
> and every their children and posterity which shall happen to be born within

the presincts (sic) thereof shall have and enjoy all liberties, franchises, immunities, capacities and abilities, of franchises and natural subjects within any of our dominions, to all intents and purposes as if they had been abiding and borne within this our realms of England or in any of our dominions.

(reproduced in Winchester, 1985, p. 165)

This charter was never repealed and yet in 1981 under the *British Nationality Act* these British people, against their very strongly expressed will, had these guaranteed rights removed from them. The act took from citizens of Britain's colonies the right of abode in the UK and their passports from that time were issued to citizens of British Dependent Territories. It is quite clear that, whatever its stated purpose, this act was passed to avoid the UK being subject to unwanted levels of immigration from Hong Kong in the run-up to that colony's reversion to China in 1997. To single out Hong Kong would have held up the British authorities, perhaps, to charges of being antipathetic towards Chinese people, so the other colonies were caught up in the legislation. Different, more favourable arrangements always applied to Gibraltar and, in 1983 after the Conflict, to the Falklands. These are the only two British colonies with predominantly white populations and there were those on St Helena who viewed the exclusion of these two territories from the legislation as being for racial reasons. This is probably untrue. Gibraltar has a position in relationship to the then European Community that could not have been squared with its citizens being denied access to part of that same Community. Regarding the Falklands, it would have been politically untenable to have fought a war on the basis of the Falklands' population being British and then continue to treat them as second class subjects. That the Saints could be so treated despite their guarantee from a British monarch, their loyalty to the British state and so completely against their will, is just a sad example of the powerlessness of a disregarded island. The effect was not just psychological. The Act cut off one easy outlet for surplus population, for only people who got work permits could take up employment in the UK; for all visits, visas were required. The *British Nationality Act* became a *cause célèbre* for the Saints; travellers to the island were harangued about it (Winchester, 1985; Ritchie, 1997; see also Wigglesworth, 1998). Pressure groups emerged; most prominently that set up by the Bishop of St Helena. The report of his commission: *St Helena, the lost county of England* (Bishop of St Helena's Commission on Citizenship, 1996) made a powerful case for rescinding the legislation. In January 1999, a British broadsheet newspaper devoted a half page to the topic, previewing an expected government white paper emanating from a conference on dependent territories held in London soon after the Labour government came into office in 1997. The article ended with a quote from the editor of the *St Helena News*, John Drummond:

There won't be a huge influx of people into the UK or off the island. It is about a sense of security, that fundamentally this is part of Britain, that Saints are fundamentally British people and they would like their citizenship back.

(*Guardian*, 6 January 1999)

When it appeared, the white paper did indeed recommend the restoration of full citizenship rights to the populations of the newly-named British Overseas Territories.

> We have examined the options carefully. We have decided that British citizenship – and so the right of abode – should be offered to those British Dependent Territory citizens who do not already enjoy it and who want to take it up.
>
> (Foreign and Commonwealth Office, 1999, paragraph 3.7)

(This offer did not relate to people living in the Sovereign Base Areas of Cyprus or British Indian Ocean Territory.) The necessary legislation is expected to be passed in 2001. Drummond's belief that few would leave the island will then be put to the test. Elsewhere he repeated his views: 'some could never leave, some might soon want to come back', but the author of the article in which this quote appeared took a different line:

> The island could be hit by a serious drain of skilled and professional people – those who could afford to head for a British home. Left behind in even greater isolation would be the humble fisherfolk and labourers.
>
> (Hawthorne, 2000, p. 43)

Isolation

One way out of St Helena's economic doldrums might be to go for tourism in a big way. The island lacks beaches, but it is scenic, has interesting old Georgian architecture and fortifications in Jamestown (Figure 10.3), has the Napoleonic legacy upon which to capitalise, including the two houses where he lived, the Briars and Longwood House, the latter now being French territory. Longwood House with its sunken garden, and Napoleon's first and now empty tomb are the premier tourist attractions. Further, St Helena is certainly 'away from it all' for those tourists attracted to the exotic. However, the St Helena government document issued to encourage outside investors has to concede that 'it is impossible to get away on a short break and a large proportion of any time away from the island is spent travelling to and from it' (Chief Secretary, 1998, p. 14). The island's infrastructure would have to be upgraded if there were to be substantially more tourism, but the major problem regarding tourism is the delivery system. St Helena has no airport, its link to the outside world is by the *R.M.S. St Helena* (6767 tonnes). This makes about four round trips per year. Typically, they go from Cardiff through the Canaries, Ascension, to St Helena; the ship then shuttles back to Ascension to take Saints there to work, goes back to St Helena and onward to Cape Town before returning. Once a year the voyage is extended to take in Tristan da Cunha, St Helena's other Dependency. The ship has 128 berths and in the late 1990s was bringing several hundred tourists a year to St Helena, a welcome contribution to the economy. However, under present arrangements, the number of tourists is capable of little expansion

Figure 10.3 Jamestown, St Helena, 1990

because of the limitations of the ship's capacity, the small number of voyages she is able to make each year and the fact that many berths have to be sold to Saints rather than tourists.

The *R.M.S. St Helena* is of immense importance to Saints. Thus when the head of Curnow Shipping, the company that manages the ship, was forced to resign in 1999 in a boardroom putsch, this was the headline in a UK-based St Helena publication: 'Stabbed in the back by traitors. [Andrew] Bell forced to quit Curnow' (*St Helena & South Atlantic News Review*, October 1999). A problem for tourists and all requiring access is that relying upon a single vessel puts communications at risk should that vessel malfunction. This is not just a theoretical possibility. A serious breakdown happened in November 1999 when *R.M.S. St Helena* suffered major engine failure whilst loaded with the island's Christmas goods. She limped into Brest in France and after a few days of specu-lation, rumour and alternative plans that monopolised conversations in the South Atlantic (the author was on Ascension Island at that time), two ships were chartered to replace the stricken vessel. Why two? Because *R.M.S. St Helena* is virtually unique in the shipping world now in having both extensive passenger and freight capability: 128 berths and 2000 tonnes. Thus vessels such as the *MV Steuart*, chartered by European and South Atlantic Lines to supply Ascension Island, do not have much passenger capacity. *MV Steuart* is licensed to carry only 17 people and that includes the crew. Ascension's passengers mainly arrive by air, or, if from St Helena, aboard the *R.M.S. St Helena*. By con-trast, normal passenger ships lack freight handling capability. So one of the emergency charters was for passengers; the other for the freight. With no airport, the only way off St Helena is by ship. Only having one ship is, as was seen, a fragile lifeline. The risk of being stranded is a disincentive to business and holiday visitors alike, as well as an inconvenience for the Saints. Christmas 1999 and the Millennium celebrations almost had to be postponed for want of goods.

To revert to tourism, cruise ships might be one answer to the delivery problem but here, again, St Helena's incredible isolation is against her. Ascension is about two days' voyage distant; Cape Town five days. The prospect of a five-day passage through the Cape rollers, the regional sea movements which strike north- or south-bound vessels on the side and make them roll up to 30 degrees, is hardly an attraction.

The only way that tourism could really take off on St Helena would be to provide the island with an airport. A site has been selected on Deadwood Plain; there is a company called St Helena Airways which has plans to run an air service; both the British and St Helena governments support the project, but no funding has been made available. The British already contribute several million pounds per annum to support the island and, having in the late 1980s and early 1990s, spent almost £32 m to replace the ship, are not willing as yet to fund an airport. The St Helena government has the will but not the money; private funding has not been attracted to the venture, given the small scale of the island and its isolation. The British government at the time of writing is searching for new civilian uses for Ascension Island. If this destination could be opened up to increased tourism, perhaps two-centre South Atlantic holidays might be possible with use of both Ascension and St Helena. Perhaps joint marketing might increase the potential number of tourists to St Helena and make an airport there less of a financial risk. In 2000, plans for rebuilding an airport were revived.

An airport on St Helena would, of course, have negative social and environmental costs: the former through the exposure of a remote society to a more brash and uncaring outside world; the latter through the loss of land and habitat. To counterbalance these negative externalities, an airport remains the only way St Helena's life and economy could be transformed. Other plans based on an improvement and increased periodicity in shipping movements could only marginally alleviate the situation (Royle, 1995).

As things stand, apart from a few yachts and cruise ships, everything and everybody comes and goes on *R.M.S. St Helena*. Passengers range from the Governor, to the rare island criminal being taken to the UK to serve a long prison term, to patients being taken to South Africa or the UK for medical treatment. And when the ship has left, the island's people and their social and welfare systems have to cope with whatever happens and it could be many days before even emergency help could be delivered by the outside world. Such exceptional isolation is a quintessentially insular phenomenon. And, of course, in the event of serious illness St Helena's isolation can become a matter of life and death. In late 1999 a St Helena girl fell ill with a rare condition. She had need of urgent treatment in a hospital with facilities simply not available in

Jamestown. She was kept alive by emergency blood transfusions from her brothers whilst radio messages were sent round the Atlantic asking for ships to call to evacuate the girl to South Africa. Fortunately one was able to come and the family, including the brothers with their depleting blood supply, all with fresh new passports, were taken to a hospital in Cape Town.

Scale and dependency

St Helena's insularity affects every aspect of its society. Being without indigenes, there is no aboriginal language or customs; everybody speaks English, if with a distinctive accent, and it has been claimed that some of the uses of language reflect the seventeenth century immigration and subsequent isolation of many of its people. The wide variety of bloodlines has been mentioned. The way this society operates is, in some respects, predicated upon scale, another impact of insularity. The island has, in many ventures, a two-tier system with the country areas having basic provisions and Jamestown providing higher order facilities. This happens with regard to health care, with six country clinics and a Jamestown hospital. Education was recently re-organised on the same model, with primary and middle school provision throughout the island and one large senior school, Prince Andrew School, just outside Jamestown. Jamestown (Figure 10.3) is the chief central place and seat of government, high order commerce, and also has the prison (which had three prisoners in July 2000). In every area the scale of the island means that the normal multi-tier hierarchical provision of services is interrupted. Off-island providers are often involved at the highest levels of provision. Patients requiring procedures not within the competence of the three doctors and one dentist with their limited range of equipment are sent off the island, usually in a more controlled way than the girl mentioned above. This would include eye surgery for example, which is not normally carried out on St Helena. Students going into tertiary education are sent off the island; as mentioned, prisoners serving long sentences are sent off the island; high order goods might have to be ordered in from off the island. Many of the highest-ranking jobs in the administration are staffed by ex-patriates. In 1990, of the senior positions such as the Chief Secretary, the Chief Medical Officer and the other two doctors, the Government Economist, the Chief Education Officer, the Headmaster of Prince Andrew School and the Bishop of St Helena, only the Chief Education Officer and the Bishop were Saints. The government system, of course, is directed by an ex-patriate British Governor with the elected Legislative Council being advisory to the governor. This domination of the top levels of the hierarchy by off-island provision causes, on occasion, resentment and, in 1998, there was an altercation in the Governor's office over development issues and hands were laid upon the Governor. The control of much of the media, especially the radio and the newspaper, by the government also causes concern (Ritchie, 1997). However, St Helena's people are not seeking alternative constitutional arrangements. Recent political clamour was for the restoration of full British citizenship, not for any distancing from Britain. There is a widespread realisation that St Helena lacks the strength of economy and the scale to function as an independent nation.

The benefits of insularity: the island as prison

One cannot end a piece on St Helena without further consideration of island's use as a prison. The one widely known fact about the island is that it was the place where Napoleon was incarcerated after his defeat at Waterloo in 1815 until his death in 1821. His being sent there after escape from the less effective island prison of Elba before Waterloo gave tiny St Helena its only place on the world stage and exemplifies perfectly one of the few indisputable positive attributes of an isolated island – its utility as a prison. In fact the British used St Helena again as a prison: to incarcerate a Zulu prince, dissident Bahraini politicians and, more significantly as a prisoner-of-war camp for Boers between 1900 and 1902 during the South African War. The imprisoned Boers approximately doubled the island's population. That St Helena was an effective prison is not in doubt; nobody escaped, although there were a few attempts at stowing away or trying to hijack boats. However, the problems of an island existence were also exacerbated at this time and the natural systems could not cope with such an increase in population. The Boers were troubled on occasion by a lack of fresh water, especially for washing; there were terrible problems with poor sanitation and many of the prisoners' deaths resulted from infectious diseases associated with sanitation problems, especially enteric fever (typhoid). The British were forced to begin to install a state of the art desalination plant at the, then, considerable cost of £5000. However, the war ended before it was completed and the remaining Boers were repatriated (Royle, 1998). St Helena, the ultimate island, is seemingly purpose-designed to serve as a prison, but even here its insular constraints made things difficult.

Conclusion: St Helena wins again

St Helena exemplifies, if in extreme form, the basic constraints of insularity: isolation, small scale, powerlessness, dependency, environmental fragility, etc. It is no surprise that the Saints and their outside rulers struggle and often fail to overcome the problems that result. Small islands are challenging places; too challenging it seems for the many island populations who, over the past 150 years or so, have abandoned them, with many more having had less extreme experience of out-migration. Small islands are also special places, where people can feel, perhaps, closer to nature or have some other romantic notion of getting away from it all. But, holidays apart, one hopes, one cannot really ever escape from the insular constraints. The hero of R.M. Ballantyne's classic island novel, *The coral island*, says:

At last we came among the Coral Islands of the Pacific and I shall never forget the delight with which I gazed – when we chanced to pass one – at the pure, white, dazzling shores and the verdant palm trees, which looked bright and beautiful in the sunshine. And often did we three long to be landed on one, imagining that we should find perfect happiness there!

(1902, p. 15)

The drama, of course, unfolded rapidly thereafter and their wish came true although 'perfect happiness' was not entrained. *The coral island* is fiction, but Ballantyne's paradox of the island of dreams versus the harsher realities of small island existence rings true – indeed this juxtaposition was explored in Chapter 1. And, of course, it can be illustrated, too, from St Helena.

On St Helena the author heard a story of an experimental wind power system of the 1980s which had blown down (though, to be fair, further investments have since been made into wind power). Much earlier, in 1909, a philanthropist, one Mr A. Mosely, opened a mackerel-canning factory:

[He] bought new fishing boats, and fishing gear. Fishermen had been engaged, the factory built, the new machinery was in order, the empty tins in thousands were there in which the mackerel were to be hermetically sealed. And there were no mackerel! Never before had there been no mackerel. The experts were unable to offer any explanation ... The factory was shut down and yet another scheme to help St Helena had failed.

(Gosse, 1938, pp. 347–8)

There are a number of other examples of development initiatives coming to naught on St Helena. On St Helena the author heard the initials 'SHWA' used to summarise these not unusual failed attempts to bring at least prosperity, if not Ballantyne's 'perfect happiness', to island life. SHWA stands for 'St Helena Wins Again'.

References

Key readings

The major historian of St Helena is Gosse, whose book takes the story up to 1938. More recent popular accounts can be seen in Bain, Ritchie and Winchester. Royle has made a number of studies of St Helena.

Aldrich, R. and Connell, J. (1998) *The last colonies*, Cambridge University Press: Cambridge.

Ashmole, P. and Ashmole, M. (2000) *St Helena and Ascension Island: a natural history*, Anthony Nelson: Oswestry.

Bain, K. (1993) *St Helena: the island, her people and their ship*, Wilton 65: York.

Ballantyne, R.M. (1902) *The coral island: a tale of the Pacific*, Blackie: London.

Bishop of St Helena's Commission on Citizenship (1996) *St Helena: the lost county of England*, Bishop of St Helena's Commission on Citizenship: St Helena.

Brown, L.C. (1981) *The land resources and agro-forestal development of St Helena*, Land Resources Development Centre, Overseas Development Administration: Surbiton.

Chief Secretary (1998) *Investor's guide to the Island of St Helena*, Office of the Chief Secretary, Government of St Helena: Jamestown.

Foreign and Commonwealth Office (1999) *A partnership for progress and prosperity: Britain and the Overseas Territories*, HMSO: London.

Gosse, P. (1938) *St Helena 1502–1938*, Cassell: London (republished 1990, Anthony Nelson: Oswestry).

Hawthorne, P. (2000) 'Will the Saints come marching home?', *Time*, 24 July, p. 48.

Keay, J. (1991) *The honourable company: a history of the English East India Company*, HarperCollins: London.

Northcliffe, G.B. (1969) 'The role of scale in locational analysis: the phormium industry on St Helena', *Journal of Tropical Geography*, 29, pp. 48–57.

Ritchie, H. (1997) *The last pink bits*, Hodder and Stoughton: London.

Royle, S.A. (1992) 'Attitudes and aspirations on St Helena in face of continued economic dependency', *Geographical Journal*, 158, pp. 31–9.

Royle, S.A. (1995) 'Economic and political prospects for the British Atlantic Dependent Territories', *Geographical Journal*, 161, pp. 307–22.

Royle, S.A. (1998) 'St Helena as a Boer prisoner of war camp, 1900–1902: information from the Alice Stopford Green papers', *Journal of Historical Geography*, 24, 1, 1998, pp. 53–68.

Royle, S.A. and the late A.B. Cross (1995) 'Wilberforce Arnold, St Helena Colonial Surgeon, 1903–1925', *Wirebird*, 12, pp. 23–8.

Royle, S.A. and the late A.B. Cross (1996) 'Health and welfare in St Helena: the contribution of W.J.J. Arnold, colonial surgeon, 1903–1925', *Health and Place*, 2, 4, pp. 239–45.

St Helena Strategic Review (1996) *The St Helena Strategic Review 1996/7 to 2000/1; the key policy and development plan for the island*, Department of Economic Development and Planning, St Helena Government: Jamestown.

Schulenburg, A.H. (2001) ' "Island of the blessed": Eden, Arcadia and the picturesque in the textualising of St Helena', *Journal of Historical Geography*, forthcoming.

Wigglesworth, A. (1998) 'Saints on the march', *Geographical*, 70, 6, pp. 57–62.

Winchester, S. (1985) *Outposts: journeys to the surviving relics of the British Empire*, Hodder and Stoughton: London (the edition used here was published in 1986 by Sceptre: London).

Index